Current Communications
IN MOLECULAR BIOLOGY
Cold Spring Harbor Laboratory Press

Polymerase Chain Reaction

Edited by

Henry A. Erlich
Cetus Corporation

Richard Gibbs
Baylor College of Medicine

Haig H. Kazazian, Jr.
Johns Hopkins University School of Medicine

A Banbury Center Meeting

POLYMERASE CHAIN REACTION

Cover: Versatility of the polymerase chain reaction (PCR). Shown is the wide size range of fragments that can be specifically amplified via PCR. (Main photograph courtesy of R. Saiki, Cetus Corporation.) The inset illustrates that many different sequences can be simultaneously amplified in a single PCR reaction. In this example, deletions were detected in the Duchenne muscular dystrophy gene. (Inset photograph courtesy of J. Chamberlain, Baylor College of Medicine.)

The articles published in this book have not been peer-reviewed. They express their authors' views, which are not necessarily endorsed by the Banbury Center or Cold Spring Harbor Laboratory.

The individual summaries contained herein should not be treated as publications or listed in bibliographies. Information contained herein can be cited as personal communication contingent upon obtaining the consent of the author. The collected work may, however, be cited as a general source of information on this topic.

All Cold Spring Harbor Laboratory Press publications may be ordered directly from Cold Spring Harbor Laboratory Press, Box 100, Cold Spring Harbor, New York 11724. (Phone: Continental U.S. except New York State 1-800-843-4388. All other locations [516] 367-8325.)

iii

Conference Participants

Norman Arnheim, Dept. of Biological Sciences, University of Southern California, Los Angeles

Thomas R. Broker, Biochemistry Dept., University of Rochester School of Medicine, New York

Jeffrey Chamberlain, Institute for Molecular Genetics, Baylor College of Medicine, Texas Medical Center, Houston

Pieter J. de Jong, Biomedical Sciences Division, Lawrence Livermore National Laboratory, Livermore, California

Anthony G. DiLella, Merck Sharp & Dohme Research Laboratories, Rahway, New Jersey

Henry A. Erlich, Human Genetics Dept., Cetus Corporation, Emeryville, California

David Gelfand, PCR Division, Cetus Corporation, Emeryville, California

Richard Gibbs, Institute for Molecular Genetics, Baylor College of Medicine, Texas Medical Center, Houston

David Ginsburg, Depts. of Internal Medicine and Human Genetics, Howard Hughes Medical Institute, University of Michigan Medical School, Ann Arbor

Dmitry Goldgaber, Dept. of Psychiatry and Behavioral Science, Health Science Center, State University of New York, Stony Brook

Lawrence Haff, Perkin-Elmer Corporation, Norwalk, Connecticut

Haig H. Kazazian, Jr., Dept. of Pediatrics, Genetics Unit, Johns Hopkins University School of Medicine, Baltimore, Maryland

Phouthone Keohavong, Center for Environmental Health Sciences, Massachusetts Institute of Technology, Cambridge

Thomas A. Kunkel, Laboratory of Molecular Genetics, National Institute of Environmental Health Sciences, Research Triangle Park, North Carolina

Ming-Sheng Lee, Division of Laboratory Medicine, M.D. Anderson Cancer Center, University of Texas, Houston

Hans Lehrach, Imperial Cancer Research Fund Laboratories, London, England

Leonard S. Lerman, Dept. of Biology, Massachusetts Institute of Technology, Cambridge

Alex F. Markham, ICI Diagnostics, Northwich, Cheshire, England

Lincoln J. McBride, Applied Biosystems, Inc., Foster City, California

Kary B. Mullis, Private Consultant, La Jolla, California

Richard M. Myers, Dept. of Physiology, University of California, San Francisco

Christian Oste, Perkin-Elmer Corporation, San Jose, California
Manuel Perucho, California Institute of Biological Research, La Jolla, California
Bernard J. Poiesz, Dept. of Medicine, Division of Hematology/ Oncology, State University of New York, Health Science Center, Syracuse
Randall K. Saiki, Dept. of Human Genetics, Cetus Corporation, Emeryville, California
Gerald Schochetman, AIDS Program, Center for Infectious Diseases, Centers for Disease Control, Atlanta, Georgia
Francis I. Smith, Dept. of Microbiology, Mt. Sinai School of Medicine, New York, New York
Oliver Smithies, Dept. of Pathology, University of North Carolina, Chapel Hill
John J. Sninsky, Dept. of Infectious Diseases, Cetus Corporation, Emeryville, California
Kenneth Tindall, Laboratory of Molecular Genetics, National Institute of Environmental Health Sciences, Research Triangle Park, North Carolina
John A. Todd, Nuffield Dept. of Surgery, John Radcliffe Hospital, Headington, Oxford, England
David Valle, Dept. of Pediatrics, Johns Hopkins University School of Medicine, Baltimore, Maryland
Albert A. van Zeeland, Dept. of Radiation Genetics and Chemical Mutagenesis, State University of Leiden, The Netherlands
James D. Watson, Cold Spring Harbor Laboratory, New York
James Weber, Marshfield Medical Research Foundation, Marshfield, Wisconsin
Michael Wigler, Cold Spring Harbor Laboratory, New York
J. Fenton Williams, Perkin-Elmer Corporation, Norwalk, Connecticut
Robert Williamson, Dept. of Bichemistry and Molecular Genetics, St. Mary's Hospital Medical School, London, England
Richard K. Wilson, Division of Biology, California Institute of Technology, Pasadena
Savio L.C. Woo, Dept. of Cell Biology, Howard Hughes Medical Institute, Baylor College of Medicine, Houston, Texas
Daniel Wu, Dept. of Molecular Biochemistry, Beckman Research Institute of the City of Hope, Duarte, California
Mark Zoller, Cold Spring Harbor Laboratory, New York

Preface

A conference on the biochemistry and multiple applications of the polymerase chain reaction was held at the Banbury Conference Center of Cold Spring Harbor during December 12–14, 1988. The meeting was conceived in discussions held earlier between Richard Gibbs and Jan Witkowski. The presentations at the meeting are summarized in this volume as extended abstracts provided by the speakers. Polymerase chain reaction was the original idea of Kary Mullis of Cetus Corporation in 1984. The technique was significantly developed by Henry Erlich, Norman Arnheim, Randy Saiki, and others in the Human Genetic Division at Cetus. It has become an extremely useful technique in molecular biology and has revolutionized the way in which a wide variety of experiments and clinical molecular genetic tests are performed. Even so, it is clear that applications of the technique are only beginning to be discovered. The actual Banbury meeting was highly informative and timely. In addition, the beautiful surroundings of the Center were enhanced by a lovely 8-inch snowstorm, which luckily did not impede transportation into and out of Cold Spring Harbor.

We are grateful to Perkin-Elmer Cetus for funding this meeting. Many thanks to Jim Watson for his support of the meeting and to Jan Witkowski, Director of the Banbury Center, and Bea Tolliver for their terrific organizational help. Katya Davey and the staff at Robertson House provided outstanding cuisine and wonderful ambience for informal discussion. We also thank Nancy Ford, Managing Director of Publications, Ralph Battey, Dorothy Brown, and Joan Ebert for expert work in compiling, editing, and publishing this volume.

H.A.E.
R.A.G.
H.H.K.

Contents

Introduction

H.A. Erlich,[1] H.H. Kazazian, Jr.,[2]
and R.A. Gibbs[3]

[1]Human Genetics Department, Cetus Corporation
Emeryville, California 94608

[2]Department of Pediatrics, Johns Hopkins University
School of Medicine, Baltimore, Maryland 21205

[3]Institute for Molecular Genetics, Baylor College of Medicine
Houston, Texas 77030

The recently developed polymerase chain reaction (PCR), a primer-directed method for the enzymatic amplification of specific DNA sequences, has transformed the way we think about molecular biology. In December of 1988, a small group of scientists all actively involved in the use of the PCR method met at the Banbury Center to discuss both the methodology and the application of PCR to a wide variety of biological problems. This volume is a compilation of presentations made at the Banbury meeting. In this short introduction, we point out some of the highlights of this volume.

One set of papers focuses on the basic method and discusses the properties of the thermostable *Taq* DNA polymerase as well as the effect of varying PCR parameters (e.g., primer concentration, Mg^{++} and dNTP concentration, enzyme concentration, and temperature profiles) on the yield and specificity of the amplification. Although for any given pair of oligonucleotide primers, an optimal set of conditions can be established, it is clear from the data presented and from the lively discussion that there is no general optimal set of conditions for all possible reactions. Thus, the reaction conditions that allow the simultaneous amplification of nine different exons of the *DMD* (dystrophin) locus ("multiplex" PCR using nine primer pairs) are quite different from those that allow the efficient and specific amplification of β-globin fragments using a single primer pair (see cover). Also addressed in these papers is the issue of fidelity—that is, the frequency of nucleotide incorporation error occurring during the PCR. The estimate of the *Taq* polymerase misincorporation rate derived from biochemical analysis of in vitro errors in replicating the β-galactosidase gene ($\sim 1 \times 10^{-4}$ nt) is in reasonable agreement with the rate (2×10^{-4} nt/cycle) estimated by the sequence analysis of many

1

individual M13 clones derived from a PCR amplification from genomic DNA. The use of modified reaction conditions (e.g., lower Mg^{++}, lower dNTPs) may lead to a lower misincorporation rate. The use of the T4 DNA polymerase, an enzyme with a very low error rate relative to *Escherichia coli* DNA polymerase I and *Taq* polymerase, appears to allow "high-fidelity" PCR but, due to its thermolability, requires addition of enzyme during each cycle. Other enzymologic properties of the *Taq* polymerase (e.g., elongation rates, exonuclease activity, and temperature optimum) are also addressed.

Another group of papers deals with the use of PCR in clinical genetics. The identification of new genetic disease mutations has been enormously simplified by using direct sequencing of amplified DNA. Similarly, the detection of known mutations (e.g., those causing β-thalassemia or phenylketonuria) can now be carried out in a simple diagnostic "dot-blot" procedure with labeled oligonucleotide probes. In addition to the oligonucleotide probe approach, diagnostic procedures involving restriction enzyme cleavage of the PCR product as well as the use of competitive oligonucleotide priming have also been applied to several diseases. The "conversion" of restriction-fragment-length polymorphism linkage markers for cystic fibrosis to PCR markers (detected either by restriction enzyme digestion or by oligonucleotide probe hybridization) for pedigree analysis and carrier identification indicated that PCR amplification can contribute to genetic diagnosis even if the disease locus and mutation have not yet been identified. The detection of various deletion mutations at the huge *DMD* (dystrophin) locus was accomplished by the coamplification of nine different exons, followed by gel electrophoresis of the PCR products with the absence of a specific fragment in males (XY/DMD) revealing the presence of a specific deletion. The potential for identifying carriers (XX/DMD) by differences in band intensities is also addressed. A completely different approach to the identification of sequence differences, both known *and* unknown, in amplified DNA is the method of denaturing gradient gel electrophoresis. Differences in the melting profiles of different sequences are reflected in the differential mobility of the amplified fragments, and the use of G+C-rich regions (a "GC clamp") at the 5′ end of the PCR primers significantly enhances the resolving power of this approach.

The analysis by PCR of highly polymorphic regions of the genome is addressed in another group of papers. Sequence variation in the HLA class II ("immune response") genes was identified initially by nucleotide sequencing and, subsequently, by the analysis of the distribution of the alleles in patient and control populations using oligonucleotide probes. For several different autoimmune diseases, specific allelic variants appear to confer susceptibility, and the sequence comparison of susceptible versus nonsusceptible alleles has revealed the importance of individual residues. Phylogenetic analysis of class II sequences amplified from a variety of primates has shown that the extensive allelic diversity observed in humans was present in an ancestral species that gave rise to the hominoid lineages (human, chimpanzee, and gorilla). The polymorphism in these genes has been used in a PCR-based genetic typing system for forensic studies. In addition, some of the general issues regarding PCR and forensic analysis, such as the effect of DNA damage on amplification, are also addressed. The ability of PCR to amplify length polymorphisms, detectable by gel electrophoresis of the PCR products, makes available for genetic analysis a large, previously inaccessible set of markers. Although the example discussed, $(dC-dA)_n \cdot (dG-dT)_n$ repeats, is represented throughout the genome, other single-locus variable number tandem repeat markers can also be amplified using flanking unique sequence primers.

Another group, discussing the detection of rare sequences, focuses on the identification of pathogenic viral sequences as well as on the analysis of rare genomic recombination events. Although conceptually quite different, the studies presented share the problem of detecting rare sequences and the concomitant problem of "contamination" or "PCR product carryover" in the analysis of amplification reactions initiated with very few templates. The clinical relevance of detecting HIV and other retroviral sequences in clinical samples is discussed, as are various approaches to address the "contamination/carryover" problem to ensure reliable pathogen diagnosis. The characterization of the human papillomavirus life cycle was also facilitated by PCR amplification.

The ability of PCR to amplify informative sequences from a single template has important implications for the analysis of recombination and, hence, for the construction of genetic maps. The coamplification of sequences from two (or more) linked loci from individual gametes represents a new and powerful approach to gene mapping. Similarly, the targeted modification of genes by a rare homologous recombination with exogenous donor sequences can now be monitored much more readily using PCR.

The contribution of PCR to the development of a fully automated, fluorescence-based sequencing is presented, as is an alternative approach to amplification in which ligation of two oligonucleotides, rather than primer-directed enzymatic synthesis, creates a new template for the following cycle. The ability of PCR to monitor gene expression following reverse transcription of an mRNA template was exploited to analyze the expressed repertoire of T-cell-receptor genes. This approach, which has yielded significant insights into the specificity of the pathogenic autoimmune response in animal models, promises to be a powerful tool in the molecular analysis of the immune system. Since the PCR primers become incorporated into the PCR product and because mismatches between the primers and initial template are tolerated, modifications of the target sequence can be introduced during amplification. The use of PCR for specific in vitro mutagenesis is discussed, and the book ends with a personal view of the origins and history of PCR technology by its inventor.

We hope that the information presented in this volume will encourage those who have not yet used PCR to consider applying it in their own special cases. Although the range of topics discussed here is already very broad, we fully expect that the number of different applications for this powerful new method will continue to grow as modifications of the basic technique are developed.

Fidelity of DNA Polymerases Used in Polymerase Chain Reactions

T.A. Kunkel and K.A. Eckert

Laboratory of Molecular Genetics, E3-01, National Institute of
Environmental Health Sciences, Research Triangle Park
North Carolina 27709

The ability of purified DNA polymerases to synthesize DNA in vitro has been valuable for a wide variety of molecular biological techniques. Nowhere is this more evident than in the case of the polymerase chain reaction (PCR). Among the potential uses for PCR reaction products are some that require accurate maintenance of the original genetic information throughout the amplification process. As DNA polymerases are not infinitely accurate reagents for duplicating DNA, an understanding of the fidelity with which they synthesize DNA in vitro is useful for a full appreciation of the utility of PCR. In this paper, we review the fidelity of several DNA polymerases currently used in molecular biology, including the large (Klenow) fragment of *Escherichia coli* DNA polymerase I (pol I), T4 DNA polymerase (T4 pol), and the DNA polymerase isolated from *Thermus aquaticus (Taq* pol). We also discuss the variables affecting accuracy and some possibilities for improving the fidelity of PCR.

Polymerase Error Rates

The simplest assay for determining the fidelity of DNA synthesis in vitro employs synthetic DNA substrates containing only one or two types of nucleotides. The error rate is defined as the ratio of noncomplementary to total nucleotides incorporated during replication of the synthetic template. Such estimates of error rate vary considerably among DNA polymerases (for review, see Loeb and Kunkel 1982), the least accurate being those enzymes that lack a proofreading exonuclease activity. For example, avian myeloblastosis virus DNA polymerase (AMV pol), a proofreading-deficient enzyme, has an error rate of 1/2000 nucleotides synthesized using the alternating copolymer poly[d(A-T)] as a template (Table 1). In contrast, synthesis by either *E. coli* pol I or T4 pol, both of which contain an associ-

5

Table 1 Base Substitution Fidelity of DNA Polymerases Used in PCR

DNA polymerase	Proofreading exonuclease activity	Error rate		
		poly [d(A-T)]	φX174 DNA	M13mp2 DNA
E. coli pol I	yes	1/10,000	1/1,000,000	1/100,000
	no[a]	—	1/50,000	1/10,000
T4 pol	yes	1/50,000	1/10,000,000	—
Taq pol	no	—	—	1/10,000
AMV pol	no	1/2,000	1/10,000	1/20,000

The data are taken from a variety of references that are reviewed in Loeb and Reyland (1987) and Kunkel and Bebenek (1988). Some values have been rounded for ease of expression. The *Taq* pol results are for synthesis at 70°C, whereas the other polymerase reactions were performed at 37°C.

[a]The polymerase reaction contained 1 mM each dNTP, which substantially inhibits the exonucleolytic proofreading activity of *E. coli* pol I (Kunkel 1988).

ated $3' \rightarrow 5'$ exonuclease activity, is considerably more accurate (1/10,000 and 1/50,000, respectively; see Table 1).

More relevant to PCR reactions, fidelity estimates can also be obtained using natural DNA substrates such as single-stranded circular DNA purified from bacteriophage φX174 or M13mp2. A DNA template that contains a phenotypically detectable mutation within a particular gene is replicated in vitro; the error rate of the polymerization reaction is estimated from the reversion frequency of the replicated viral DNA. This reversion frequency is quantitated as the proportion of wild-type (revertant) plaques to the total number of phage plaques obtained after transfection of bacterial cells with the copied DNA product.

In these assays, polymerization by pol I, *Taq* pol, or AMV pol in the absence of a functional proofreading exonuclease activity creates about one single-base substitution error for each 10,000–50,000 bases polymerized (Table 1), as measured by reversion of a TAG codon (φX174) or a TGA codon (M13mp2). An equivalent measurement in the absence of proofreading has not been made for T4 pol, since it has not yet been possible to incapacitate its powerful proofreading activity effectively. The base substitution fidelity of polymerases is higher when a proofreading exonuclease is operating to remove misinserted bases (e.g., pol I in Table 1). In some instances, proofreading can improve accuracy more than 100-fold (for review, see Kunkel 1988).

The assays described above are limited in that they monitor only single-base substitution errors that are capable of reverting the TAG and TGA nonsense codons. A broader description of the many possible errors that could occur during PCR reactions with a larger DNA target sequence can be obtained using the M13mp2 forward mutation assay (Kunkel 1985). This assay scores for loss of a nonessential gene function, *lacZα*-complementation, and therefore detects a wide variety of mutations. Over 200 different base substitution errors at 110 different sites can be scored within the 250-base *lacZα* target sequence, along with frameshift mutations, deletions, and more complex errors. A gapped M13mp2 DNA molecule in which the gap contains the target sequence serves as a primer-template for polymerization reactions. Accurate in vitro DNA synthesis results in dark-blue M13mp2 plaques, whereas polymerase errors during gap-filling synthesis result in M13mp2 plaques having light-blue or no (white) color due to decreased α-complementation of β-galactosidase activity. The M13mp2 *lacZα* mutant frequency determines the polymerase error rate and is defined as the number of light-blue and white plaques relative to the total number of plaques scored. Following DNA sequence analysis of the confirmed mutants, error rates can be calculated for each specific class of polymerase error.

The fidelity of various polymerases, including those of interest for PCR, in the M13mp2 forward mutation assay is shown in Table 2. The least accurate DNA synthesis is catalyzed by the proofreading-deficient *Taq* pol, which at 70°C produces both base substitution and frameshift errors (primarily minus-1-base errors). Using 1 mM dNTPs to inhibit the relatively weak exonuclease activity associated with the Klenow polymerase, Klenow polymerase is about as accurate as AMV pol, both of which are two- to fourfold more accurate than the *Taq* pol (Table 2). The most accurate enzyme both at high and low dNTP concentrations is T4 pol (Table 2; data not shown), consistent with the presence of a highly active proofreading activity. These T4 pol results should be considered preliminary, since they reflect limited data and yield a mutant frequency only slightly above the spontaneous background.

Parameters That Affect Fidelity
The fidelity of DNA synthesis is dependent on the polymerase (exonuclease) being employed, the exact error under consideration, and the DNA sequence of the template-primer. For each of

Table 2 Fidelity of DNA Polymerases in the M13mp2 Forward Mutation Assay

DNA polymerase	Mutant frequency[a] ($\times 10^{-4}$)	Error rate[b] base substitutions	frameshifts
Klenow	37	1/29,000	1/65,000
T4 pol	7.4	1/160,000	1/280,000
Taq pol	110	1/9,000	1/41,000
AMV pol	31	1/37,000	1/87,000

The results for the pol I and *Taq* pol are from Tindall and Kunkel (1988) and for AMV pol are from Roberts et al. (1989). The T4 pol results are unpublished data (M.P. Fitzgerald and T.A. Kunkel). Polymerization reactions were done using 1 mM each dNTP at 37°C, except for *Taq* pol, where synthesis was carried out at 70°C.

[a]A spontaneous background mutant frequency for uncopied DNA of 6.7×10^{-4} has been subtracted.

[b]Expressed as errors per detectable polymerized base (for review, see Kunkel and Bebenek 1988) based on the DNA sequence analysis of the following number of mutants: Klenow = 24; T4 pol = 16; *Taq* pol = 42; and AMV pol = 66.

these variables, fluctuations of greater than tenfold have been reported; when combined, the total variation can be enormous. For example, the single-base substitution error rate of HIV-1 RTase is only 1/70 at one position in the *lacZα* gene (Roberts et al. 1988), and yet it is 1/830,000 for calf thymus DNA polymerase-α at one position in φX174 DNA (Reyland and Loeb 1987). Neither enzyme has a proofreading activity, yet the difference is >10,000-fold. Substantial differences among enzymes in frameshift error rates and in the proficiency of proofreading have been described previously (Kunkel 1988; Kunkel and Bebenek 1988).

Although the explanations for these effects on fidelity have not yet been fully described, studies in several laboratories have demonstrated that the fidelity of DNA synthesis in vitro is influenced by reaction conditions. Of critical importance are both the relative and the absolute concentrations of the dNTP substrates. Relative deoxynucleotide trisphosphate pool imbalances can be either mutagenic or antimutagenic, depending on the pool bias and the error(s) being considered. This reflects the relative probability that a polymerase will bind an incorrect versus a correct dNTP. The probability of the initial misinsertion event also depends on the absolute dNTP concentration, as does the probability of incorporating the next correct nucleotide onto the mispair (extension). Unextended misinsertions should

8

be corrected efficiently if the polymerase contains an associated proofreading exonuclease. For the *Taq* pol, which lacks a proofreading exonuclease, unextended errors will be lost in PCR reactions because they do not yield full-length products for further amplification. Thus, maximum fidelity can be achieved by performing reactions with a low but equal concentration of all four dNTPs.

The fidelity of DNA synthesis may also be affected by the processivity of polymerization (for review, see Kunkel and Bebenek 1988). We have previously demonstrated that eukaryotic DNA polymerase-α (pol α) is more accurate than is DNA polymerase-β (pol β). Neither contains a proofreading exonuclease, but pol α is more processive than pol β, i.e., it adds more nucleotides per association with a template-primer. Furthermore, when reaction conditions were used that are known to increase the processivity of pol β, this polymerase became more accurate. Extrapolating these results to the polymerases used for PCR, fidelity may be improved beyond the values shown in Tables 1 and 2 by establishing highly processive reaction conditions. Parameters that are known to alter the processivity of polymerization include the pH of the reaction, the type and amount of divalent metal cation used to activate the polymerase, the dNTP concentration, and the ionic strength of the reaction. Finally, the influence of reaction temperature on fidelity warrants thorough examination. The fidelity of *Taq* pol varied less than twofold for reactions performed at 55°C versus 70°C (Tindall and Kunkel 1988). We are now measuring the fidelity of both *Taq* pol and Klenow over a wider temperature range. The combination of optimal in vitro dNTP concentration, processivity, and temperature conditions may enhance the fidelity of these DNA polymerases, thus generating more useful reagents for PCR.

REFERENCES

Kunkel, T.A. 1985. The mutational specificity of DNA polymerase-β during *in vitro* DNA synthesis. Production of frameshift, base substitution and deletion mutations. *J. Biol. Chem.* **260:** 5787.

――――. 1988. Exonucleolytic proofreading. *Cell* **53:** 837.

Kunkel, T.A. and K. Bebenek. 1988. Recent studies of the fidelity of DNA synthesis. *Biochim. Biophys. Acta* **951:** 1.

Loeb, L.A. and T.A. Kunkel. 1982. Fidelity of DNA synthesis. *Annu. Rev. Biochem.* **52:** 429.

Loeb, L.A. and M.E. Reyland. 1987. Fidelity of DNA synthesis. In *Nucleic acids and molecular biology* (ed. F. Eckstein and D.M.J. Lilley), vol. 1, p. 157. Springer-Verlag, Berlin.

Reyland, M.E. and L.A. Loeb. 1987. On the fidelity of DNA replication: Isolation of high fidelity DNA polymerase-primase complexes by immunoaffinity chromatography. *J. Biol. Chem.* **262:** 10824.

Roberts, J.D., K. Bebenek, and T.A. Kunkel. 1988. The accuracy of reverse transcriptase from HIV-1. *Science* **242:**1171.

Roberts, J.D., B.D. Preston, L.A. Johnston, A. Soni, L.A. Loeb, and T.A. Kunkel. 1989. The fidelity of two retroviral reverse transcriptases during DNA-dependent DNA synthesis *in vitro. Mol. Cell. Biol.* **9:** 469.

Tindall, K.R. and T.A. Kunkel. 1988. The fidelity of DNA synthesis by the *Thermus aquaticus* DNA polymerase. *Biochemistry* **27:** 6008.

Thermus aquaticus DNA Polymerase

D.H. Gelfand

PCR Division, Cetus Corporation
Emeryville, California 94608

The polymerase chain reaction (PCR) method for amplifying se-
lectively discrete segments of DNA has found wide-spread applications in molecular biology, due in part to the substitution
of a thermostable DNA polymerase isolated from *Thermus
aquaticus* (*Taq*) (Saiki et al. 1988) for the previously used *Escherichia coli* DNA polymerase I–Klenow fragment (Saiki et al.
1985; Mullis and Faloona 1987). Since *Taq* DNA polymerase
can withstand repeated exposure to the high temperatures
(94–95°C) (Saiki et al. 1988) required for strand separation, the
tedium and frequent rebellion resulting from having to add
polymerase I–Klenow fragment after each cycle is minimized.

Taq strain YT1, a thermophilic, eubacterial microorganism
capable of growth at 70–75°C, was isolated from a hot spring in
Yellowstone National Park and first described 20 years ago
(Brock and Freeze 1969). DNA polymerase activities with an
estimated molecular mass of 60–68 kD and an inferred specific
activity of 2000–8000 units/mg have been isolated previously
from this organism (Chien et al. 1976; Kaledin et al. 1980). In
contrast, we have isolated a DNA polymerase activity from *Taq*
with a specific activity of 200,000 units/mg (S. Stoffel, in prep.),
that migrates on SDS-polyacrylamide gel electrophoresis
slightly faster than phosphorylase B (97.3 kD) and which has
an inferred molecular weight (based on DNA sequence information) of 93,910 (Lawyer et al. 1989).

Temperature Effects

As observed for several DNA polymerase activities isolated
from thermophilic microorganisms, 94-kD *Taq* DNA polymerase has a relatively high temperature optimum (T_{opt}) for DNA
synthesis. Depending on the nature of the DNA template, we
have found an apparent T_{opt} of 75–80°C with an activity approaching 150 nt/sec/enzyme molecule. Innis et al. (1988) have
reported highly processive synthesis properties and an exten-

11

sion rate of >60 nt/sec at 70°C with *Taq* DNA polymerase for a GC-rich 30-mer primer on M13 and significant extension activity at 55°C (24 nt/sec). Even at lower temperatures, *Taq* DNA polymerase has extension activities of approximately 0.25 and 1.5 nt/sec at 22°C and 37°C, respectively. At lower temperatures, there is a marked attenuation in the apparent processivity of *Taq* DNA polymerase. This could reflect an impaired ability of *Taq* DNA polymerase to extend through regions of local intramolecular secondary structure on the template strand or a change in the ratio of the forward rate constant to the dissociation constant. Very little DNA synthesis is seen at very high temperatures (>90°C). DNA synthesis at higher temperatures in vitro may be limited by the stability of the primer or priming strand and the template strand duplex.

Although *Taq* DNA polymerase has a very limited ability to synthesize DNA above 90°C, the enzyme is relatively stable to and is not denatured irreversibly by exposure to high temperature. In a PCR mixture, *Taq* DNA polymerase retains 50% of its activity after about 130 minutes, 40 minutes, and 5–6 minutes at 92.5°C, 95°C, and 97.5°C, respectively (R. Watson, pers. comm.). Preliminary results indicate retention of 65% activity after a 50-cycle PCR when the upper limit temperature (in tube) is 95°C for 20 seconds in each cycle. This is consistent with the static stability values.

Mg++/dNTPs

Taq DNA polymerase activity is sensitive to the concentration of magnesium ion, as well as to the nature and concentration of monovalent ions (D.H. Gelfand, in prep.). Using minimally activated salmon sperm DNA as template in a standard 10-minute assay (Lawyer et al. 1989), 2.0 mM magnesium chloride maximally stimulates *Taq* polymerase activity at 0.7–0.8 mM total deoxynucleoside triphosphate (dNTP). Higher concentrations of Mg^{++} are inhibitory, with 40–50% inhibition at 10 mM $MgCl_2$. Since dNTP can bind Mg^{++}, the precise magnesium concentration that is required to activate the enzyme maximally is dependent on the dNTP concentration. In addition, at the correspondingly optimal magnesium concentration, the synthesis rate of *Taq* polymerase decreases by as much as 20–30% as the total dNTP concentration is increased to 4–6 mM. This observation may result from substrate inhibition.

Low, balanced concentrations of dNTPs have been observed to give satisfactory yields of PCR product, to result frequently in improved specificity, to facilitate labeling of PCR products with radioactive or biotinylated precursors, and to contribute to increased fidelity of *Taq* DNA polymerase (see below).   In a 100 μl PCR with 40 μM each dNTP, there are sufficient nucleoside triphosphates to yield 2.6 μg of DNA when only half of the available dNTPs are incorporated into DNA. It is likely that very low dNTP concentrations may adversely affect the processivity of *Taq* DNA polymerase. Furthermore, the precise concentration of free and enzyme-bound magnesium may affect the processivity of *Taq* polymerase, as has been inferred for calf thymus DNA polymerases α and δ (Sabatino et al. 1988).

Associated Nuclease Activity/Fidelity

Purified 94-kD *Taq* DNA polymerase does not contain an inherent 3'-5' exonuclease activity (Tindall and Kunkel 1988; S. Stoffel, in prep.). Single nucleotide incorporation/misincorporation, biochemical fidelity measurements have indicated that the ability of "nonproofreading" DNA polymerases to misincorporate a deoxynucleoside triphosphate is determined critically by the concentration of that triphosphate (Mendelman et al. 1989). Accordingly, a model of "K_m (Michaelis constant) or V_{max} (maximum velocity) discrimination" has been advanced to suggest mechanisms by which nonproofreading DNA polymerases may achieve high fidelity. Similar data have been obtained with regard to extension of a mismatched primer/template (Petruska et al. 1988). Although not yet measured kinetically, *Taq* DNA polymerase appears to extend a mismatched primer/template significantly less efficiently than a correct primer/template (see Fig. 2 in Innis et al. 1988). It is not known if *Taq* contains a separate 3'-5' exonuclease activity that may be associated with the polymerase in vivo. Since the purification protocol for native *Taq* DNA polymerase (D.H. Gelfand, in prep.) was intended to yield a single polypeptide chain enzyme, we could have failed to detect an *E. coli* polymerase III "ε-like" associated subunit.

Taq DNA polymerase has a DNA synthesis-dependent, strand replacement, 5'-3' exonuclease activity. There is little, if any, degradation of a 5' ^{32}P-labeled oligodeoxynucleoside, either as single-stranded DNA or where annealed to an M13 template. Furthermore, the presence of a "blocking," annealed, nonextendable oligodeoxynucleoside "primer" (3' phosphorylated during synthesis) fails to attenuate incorporation from a 3' OH terminated upstream primer. There is little, if any, dis-

13

placed "blocking primer," and the products of exonuclease action are primarily deoxynucleoside monophosphate (85%) and dinucleoside phosphate (15%) (S. Stoffel, unpubl.).

Monovalent Salts and Inhibitors

Modest concentrations of potassium chloride (KCl) stimulate the synthesis rate of *Taq* polymerase by 50–60%, with an apparent optimum at 50 mM. Higher KCl concentrations begin to inhibit activity, and no significant activity is observed in a DNA-sequencing reaction at ≥75 mM KCl (Innis et al. 1988) or in a 10-minute incorporation assay at >200 mM KCl. The addition of either 50 mM ammonium chloride or ammonium acetate or sodium chloride to a *Taq* DNA polymerase activity assay results in mild inhibition, no effect, or slight stimulation (25–30%), respectively.

Low concentrations of urea, dimethylsulfoxide (DMSO), dimethylformamide (DMF), or formamide have no effect on *Taq* polymerase's incorporation activity (Table 1). The presence of 10% DMSO (used previously in Klenow-mediated PCR) (Scharf et al. 1986) in a 70°C *Taq* polymerase activity assay inhibits

Table 1 Inhibitor Effects on *Taq* Polymerase I Activity

Inhibitor	Concentration	% Activity*
Ethanol	≤3%	100
	10%	110
Urea	≤0.5 M	100
	1.0 M	118
	1.5 M	107
	2.0 M	82
DMSO	≤1%	100
	10%	53
	20%	11
DMF	≤5%	100
	10%	82
	20%	17
Formamide	≤10%	100
	15%	86
	20%	39
SDS	0.001%	105
	0.01%	10
	0.1%	≤0.1

*dNTP incorporation activity at 70°C with salmon sperm DNA/10 min.

DNA synthesis by 50%. Although several investigators have observed that inclusion of 10% DMSO facilitates certain PCR assays, it is not clear which parameters of PCR are affected. The presence of DMSO may affect the T_m of the primers, the thermal activity profile of *Taq* DNA polymerase, and/or the degree of product strand separation achieved at a particular "denaturation" or upper-limit temperature. Curiously, 10% ethanol fails to inhibit *Taq* activity and 1.0 M urea stimulates *Taq* activity. These effects on incorporation activity may not reflect the degree to which these agents affect the PCR. For example, urea at 0.5 M completely inhibits a PCR assay (C.A. Chang, pers. comm.). Finally, the inhibitory effects of low concentrations of SDS can be completely reversed by high concentrations of certain nonionic detergents (e.g., Tween 20 and Nonidet P40 [NP40]). Thus, 0.5% each Tween 20/NP40 instantaneously reverses the inhibitory effects of 0.01% SDS, and 0.1% each Tween 20/NP40 completely reverses the inhibitory effects of 0.01% SDS in the presence of DNA and Mg^{++} (no dNTP) after 40 minutes at 37°C (S. Stoffel, in prep.).

Similarity to *E. coli* Polymerase I and Future Directions
Taq DNA polymerase shows considerable amino acid sequence similarity to *E. coli* polymerase I (Lawyer et al. 1989). Significant similarity is observed in regions of the amino-terminal one-third of the two enzymes. This domain is known to contain the 5'-3' exonuclease domain of polymerase I. Of particular interest is the observation that all of the sites known to be critical for the 5'-3' exonuclease activity of *E. coli* polymerase I are perfectly conserved in *Taq* DNA polymerase. Significant identity is also observed for regions of polymerase I that are involved in 3' OH primer interaction, dNTP binding, and DNA template binding. In contrast, no meaningful alignment was obtained for the 3'-5' exonuclease domain of *E. coli* polymerase I, possibly accounting for the observed lack of 3'-5' exonuclease activity in *Taq* DNA polymerase.
Aspects of DNA synthesis at high temperature, enzymatic, biochemical, and structural properties of *Taq* and other thermostable DNA polymerases, as well as the identification and characterization of replication accessory proteins are all challenging areas for further investigation. To what degree is *Taq* DNA polymerase capable of efficient displacement synthesis at a replication fork? If only limited synthesis occurs under such conditions, are there *Thermus* "helicases" that facilitate

efficient displacement synthesis? The ability to catalyze processive displacement synthesis could ameliorate one of the factors that may contribute to the "plateau effect" and limit the final amount of specific product accumulation in a PCR. What are the structural features that contribute to the thermostability of *Taq* DNA polymerase? Since the regions and specific amino acid residues of *E. coli* polymerase I that are known to be critically important for 5'-3' exonuclease activity, dNTP binding, primer and template interaction are remarkably conserved in *Taq* DNA polymerase, other portions of the enzyme must critically determine thermostability and activity at high temperature. A comprehensive understanding of the enzymatic, biochemical, and structural properties of thermostable DNA polymerases is likely to lead to an improved ability to generate very large products, as well as to improved specificity, final yield of desired product, and enhanced sensitivity of detection of rare targets in a PCR.

ACKNOWLEDGMENTS
I thank Chu-An Chang, Susanne Stoffel, and Robert Watson for permission to cite their data prior to publication; Corey Levenson, Laurie Goda, and Dragan Spasic for the synthesis of many oligonucleotides; Lynn Mendelman for providing a preprint prior to publication; Will Bloch and members of the Cetus PCR group for stimulating discussions, thoughtful advice, and suggestions. I also thank Tom White and Jeff Price for their interest, patience, wisdom, and support of these efforts.

REFERENCES
Brock, T.D. and H. Freeze. 1969. *Thermus aquaticus* gen.n. and sp.n. a non-C sporulating extreme thermophile. *J. Bacteriol.* **98:** 289.
Chien, A., D.B. Edgar, and J.M. Trela. 1976. Deoxyribonucleic acid polymerase from the extreme thermophile *Thermus aquaticus. J. Bacteriol.* **127:** 1550.
Innis, M.A., K.B. Myambo, D.H. Gelfand, and M.A.D. Brow. 1988. DNA sequencing with *Thermus aquaticus* DNA polymerase and direct sequencing of PCR-amplified DNA. *Proc. Natl. Acad. Sci.* **85:** 9436.
Kaledin, A.S., A.G. Slyusarenko, and S.I. Gorodetskii. 1980. Isolation and properties of DNA polymerase from extremely thermophilic bacterium *Thermus aquaticus* YT1. *Biokhimiya* **45:** 644.
Lawyer, F.C., S. Stoffel, R.K. Saiki, K. Myambo, R. Drummond, and D.H. Gelfand. 1989. Isolation, characterization, and expression in *Escherichia coli* of the DNA polymerase gene from *Thermus aquaticus. J. Biol. Chem.* **264:** 6427.

Mendelman, L.V., M.S. Boosalis, J. Petruska, and M.F. Goodman. 1989. Nearest neighbor influences on DNA polymerase insertion fidelity. *J. Biol. Chem.* (in press).

Mullis, K.B. and F. Faloona. 1987. Specific synthesis of DNA *in vitro* via a polymerase catalyzed chain reaction. *Methods Enzymol.* **155:** 335.

Petruska, J., M.F. Goodman, M.S. Boosalis, L.C. Sowers, C. Cheong, and I. Tinoco, Jr. 1988. Comparison between DNA melting thermodynamics and DNA polymerase fidelity. *Proc. Natl. Acad. Sci.* **85:** 6252.

Sabatino, R.D., T.W. Myers, R.A. Bambara, O. Kwon-Shin, R.L. Marraccino, and P.H. Frickey. 1988. Calf thymus DNA polymerases α and δ are capable of highly processive DNA synthesis. *Biochemistry* **27:** 2998.

Saiki, R.K., D.H. Gelfand, S. Stoffel, S.J. Scharf, R. Higuchi, G.T. Horn, and H.A. Erlich. 1988. Primer-directed enzymatic amplification of DNA with a thermostable DNA polymerase. *Science* **239:** 487.

Saiki, R.K., S. Scharf, F. Faloona, K.B. Mullis, G.T. Horn, H.A. Erlich, and N. Arnheim. 1985. Enzymatic amplification of β-globin genomic sequences and restriction site analysis for diagnosis of sickle cell anemia. *Science* **230:** 1350.

Scharf, S.J., G.T. Horn, and H.A. Erlich. 1986. Direct cloning and sequence analysis of enzymatically amplified genomic sequences. *Science* **233:** 1076.

Tindall, K.R. and T.A. Kunkel. 1988. Fidelity of DNA synthesis of the *Thermus aquaticus* DNA polymerase. *Biochemistry* **27:** 6008.

Fidelity of DNA Amplification In Vitro

P. Keohavong and W.G. Thilly

MIT Center for Environmental Health Sciences, Massachusetts
Institute of Technology, Cambridge, Massachusetts 02139

The polymerase chain reaction (PCR) has accelerated progress
in the analysis of point mutations (Saiki et al. 1985; Mullis and
Faloona 1987; Wrischnik et al. 1987; Engelke et al. 1988). The
key to its success is the ability to perform primer extension to
copy DNA templates using DNA polymerases in vitro. DNA
polymerases themselves make errors while copying DNA templates in vitro, with error rates varying with the type of the
DNA polymerase and in vitro conditions (Loeb and Kunkel
1982; Kunkel et al. 1984). In PCR, each individual mutant sequence generated by a DNA polymerase is amplified along with
the wild-type sequences, and the fraction of mutant sequences
relative to wild-type sequences increases linearly with the
number of amplification cycles. Depending on the DNA sequence, DNA polymerase, amplification conditions, and the
number of DNA doublings desired, the mutant fraction (MF)
may become quite significant. In general, starting with a duplex sequence of bases (b), PCR conditions yielding a mutation
rate of f mutations per base and doubling and requiring doublings (d) will result in the final ratio of mutant to nonmutant
duplexes of $MF = b \times f \times d$. As an example, according to the
reported mutation rate, $f = 2 \times 10^{-4}$ for *Thermus aquaticus*
(*Taq*) DNA polymerase (Saiki et al. 1988), a millionfold amplification ($d = 20$) of a 100-bp duplex ($b = 200$) will yield a mutant
fraction of 0.8. The actual distribution of mutant base pairs
among specific DNA sequences and among possible homoduplexes and heteroduplexes is quite complex. For instance,
with $f = 10^{-4}$, a multifold amplification would be expected to
create a condition in which only a small fraction of the amplified products contains the original sequence to be studied.
Fortunately, PCR products can be used for direct sequencing
analysis and can produce unambiguous results, since the mutations are distributed over a number of base pairs such that the
original base at a particular position is still numerically predominant in the final mixture.

To study the naturally occurring mutations that would be present at a frequency of from 10^{-7} to 10^{-5} (Thilly 1985) in cell populations, an f of 10^{-4} would obscure such a small mutant fraction. Thus, to improve the overall fidelity of PCR, the fidelity of DNA amplification in vitro was studied using several types of DNA polymerases. Moreover, denaturing gradient gel electrophoresis (DGGE) (Fischer and Lerman 1983; Lerman et al. 1984) was employed to separate duplexes containing only wild-type sequence from heteroduplexes containing mutant base pairs in either strand (Thilly 1985; Myers and Maniatis 1986). This technique permitted the separation of single base-pair substitutions occurring in exon 3 of the hypoxanthine-guanine phosphoribosyltransferase (*HPRT*) gene (Cariello et al. 1988). Since heteroduplexes may form from all errors occurring during PCR, this approach has been applied to study the fidelity of DNA amplification in vitro (P. Keohavong and W.G. Thilly, in prep.).

The 184-bp exon 3 sequence of the human *HPRT* gene was amplified from genomic DNA using T4 DNA polymerase (Keohavong et al. 1988) and *Taq* DNA polymerase (Saiki et al. 1988). The amplification was carried out for 10^5-, 10^8-, and 10^{11}-fold increases using end-labeled primers, and the radioactive DNA was analyzed as heteroduplex forms by DGGE. Figure 1 shows the patterns of mutant sequences induced by T4 and *Taq* DNA polymerases. Each DNA polymerase produced errors resulting in an accumulation of mutant sequences that were separated from correctly amplified sequences (wild type) as specific distributions in lower concentrations of denaturant in the gel. Densitometer estimates of the amount of mutant sequences relative to that of the wild-type sequence after 10^8-fold amplification showed that the error per nucleotide per duplication was 3×10^{-6} for T4 DNA polymerase and 6.5×10^{-5} for *Taq* DNA polymerase. These results indicate that T4 DNA polymerase carried out the DNA amplification in vitro with a fidelity higher than that of *Taq* DNA polymerase.

The most frequent mutant sequences generated by each of the DNA polymerases after 10^8-fold amplification, as seen in Figure 1, represented from 1% to 5% of the wild type and contained between 1×10^9 to 3×10^9 copies by comparison with internal standards (not shown). These sequences were individually purified from the gel to be reamplified further. Then, after separation by a second DGGE and purification, the nature of the mutations was determined by sequencing each individual

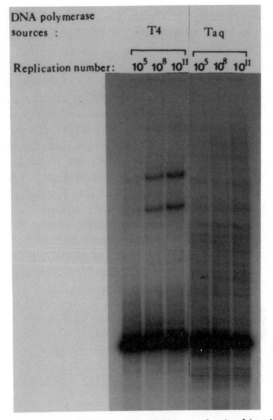

Figure 1 Analysis by DGGE of the DNA synthesized in vitro by PCR. The exon 3 sequence of the human *HPRT* gene was obtained by a 10^6- and 10^8-fold amplification from genomic DNA using ^{32}P end-labeled primers and catalyzed by the DNA polymerases indicated. Equal amounts of radioactivity (10^5 cpm) from each amplified DNA was boiled and reannealed to transform mutant sequences into heteroduplex forms. The DNA was separated by electrophoresis on a 12.5% polyacrylamide gel containing 16–30% of denaturant concentrations (urea plus formamide). The correctly amplified DNA (wild type) focused at 24% denaturant, and the polymerase-induced mutant sequences focused at lower denaturant concentrations.

mutant sequence. In this manner, the nature of the predominant mutations generated by T4 DNA polymerase and *Taq* DNA polymerase was determined. The predominant mutation generated by T4 DNA polymerase was a G-C to A-T transition as can be seen in Figure 1 with a tenfold amplification. The complementary strand of mutants with wild type yield two

21

heteroduplexes of equal intensity. Five of the most frequently observed mutations generated by *Taq* DNA polymerase all corresponded to A-T to G-C transitions.

Both the mutation rate and the nature of the mutations found for *Taq* DNA polymerase, using the protocol described here, are consistent with those obtained by cloning the amplified products and by clone-by-clone sequencing to determine the nature of the mutations (Saiki et al. 1988). Furthermore, a mutation rate of 1.1 x 10^{-4} base-pair substitutions per nucleotide synthesized, and the predominant T to C transitions have been reported by a single-round *Taq*-directed DNA synthesis using the M13mp2 assay (Tindall and Kunkel 1988). The error rates determined by three different protocols are thus in marked agreement. Because cloning of DNA segments is not required, it appears that the combination of PCR and DGGE can be used as a tool to analyze the fidelity of DNA amplification protocols. In addition, this approach facilitates studies of DNA polymerases' fidelity and auxiliary factors in vitro.

ACKNOWLEDGMENTS

This work was supported by grants from the U.S. National Institutes of Environmental Health Sciences (grant 1-P42-ES04675, 5-P01-ES00597, I-P50-ES03926-05) and the Department of the Environment (grant DE-FGO2-86-ER60448).

REFERENCES

Cariello, N.F., A. Kat, W.G. Thilly, and P. Keohavong. 1988. Resolution of a missense mutant in human genomic DNA by denaturing gradient gel electrophoresis and direct sequencing using *in vitro* DNA amplification: HPRT-$_{Munich}$. *Am. J. Hum. Genet.* **42**: 726.

Engelke, D.R., P.A. Hoener, and F.S. Collins. 1988. Direct sequencing of enzymatically amplified human genomic DNA. *Proc. Natl. Acad. Sci.* **85**: 544.

Fischer, S.G. and L.S. Lerman. 1983. DNA fragments differing by single base-pair substitutions separated in denaturing gradient gels: Correspondence with melting theory. *Proc. Natl. Acad. Sci.* **80**: 1579.

Keohavong, P., A. Kat, N.F. Cariello, and W.G. Thilly. 1988. DNA amplification *in vitro* using T4 DNA polymerase. *DNA* **7**: 63.

Kunkel, T.A., L.A. Loeb, and M.F. Goodman. 1984. On the fidelity of DNA replication. *J. Biol. Chem.* **259**: 1539.

Lerman, L.S., S.G. Fischer, I. Hurley, K. Silverstein, and N. Lumelsky. 1984. Sequenced-determined DNA separations. *Annu. Rev. Biophys. Bioeng.* **13**: 399.

Loeb, L.A. and T.A. Kunkel. 1982. Fidelity of DNA synthesis. *Annu. Rev. Biochem.* **52**: 429.

Mullis, K.B. and F.A. Faloona. 1987. Specific synthesis of DNA *in vitro* via a polymerase catalyzed chain reaction. *Methods Enzymol.* **155**: 335.

Myers, R.M. and T. Maniatis. 1986. Recent advances in the development of methods for detecting single-base substitutions associated with human genetic diseases. *Cold Spring Harbor Symp. Quant. Biol.* **51**: 275.

Saiki, R.K., S. Scharf, F. Faloona, K.B. Mullis, G.T. Horn, H.A. Erlich, and N. Arnheim. 1985. Enzymatic amplification of beta-globin genomic sequences and restriction sites analysis for analysis of sickle cell anemia. *Science* **230**: 1350.

Saiki, R.K., D.H. Gelfand, S. Stoffen, S.H. Scharf, R. Higuchi, G.T. Horn, K.B. Mullis, and H.A. Erlich. 1988. Primer-directed enzymatic amplification of DNA with a thermostable DNA polymerase. *Science* **239**: 487.

Thilly, W.G. 1985. Potential use of denaturing gradient gel electrophoresis in obtaining mutational spectra from human cells. In *Carcinogenesis: The role of chemicals and radiation in the etiology of cancer* (ed. E. Huberman and S.H. Barr), p. 511. Raven Press, New York.

Tindall, K.R. and T.A. Kunkel. 1988. Fidelity of DNA synthesis by the *Thermus aquaticus* DNA polymerase. *Biochemistry* **27**: 6008.

Wrischnik, L.A., R.G. Higuchi, M. Stoneking, H.E. Erlich, N. Arnheim, and A.C. Wilson. 1987. Length mutations in human mitochondrial DNA. *Nucleic Acids Res.* **15**: 529.

Optimization of the Polymerase Chain Reaction

R.K. Saiki

Department of Human Genetics, Cetus Corporation
Emeryville, California 94608

One aspect of the polymerase chain reaction (PCR) that has become evident over the course of the past few years is that it is a very dynamic biochemical reaction. Although the basic principles are elegantly simple and straightforward, the reaction itself involves complex kinetic interactions between template, product, primer, nucleotide triphosphates, and enzymes that change throughout the course of the reaction. Despite this complexity, PCR with the thermostable *Taq* DNA polymerase is a remarkably robust technique that works well with most amplification targets. There are, however, adjustments that can be made to some of the reaction parameters that in some cases will dramatically improve specificity and yield. These adjustments include alteration of the reaction buffer, particularly $MgCl_2$, primer concentration, dNTP concentration, enzyme concentration, annealing time and temperature, and extension time and temperature.

Reaction Buffer

In addition to the DNA sample to be amplified, a PCR reaction mix includes buffer, deoxyribonucleotide triphosphates, a pair of primers specific for the target sequence, and the thermostable *Taq* DNA polymerase. The buffer for PCR with *Taq* polymerase contains 50 mM KCl, 10 mM Tris (pH 8.4), 1.5 mM $MgCl_2$, and 100 µg/ml gelatin. Both KCl and Tris can usually be lowered without major effect. Some investigators have reported slightly better results with lower KCl concentrations (Innis et al. 1988). Gelatin is used instead of bovine serum albumin, because it is less likely to coagulate during the denaturation step and is readily sterilized in an autoclave. However, carrier protein does not always seem to be necessary, and the gelatin may be left out in situations where extraneous protein is undesired (Innis et al. 1988). The component with the most dramatic effect on specificity is $MgCl_2$; optimal levels will vary

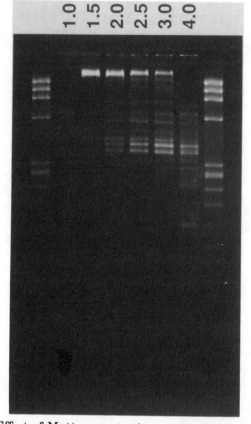

Figure 1 Effect of Mg^{++} concentration on specificity. Samples with different concentrations of MgCl$_2$ in standard buffer with 1 μM each primer and 200 μM each dNTP were subjected to 30 cycles of amplification. The primers specify a 1780-bp fragment that encompasses the entire β-globin gene. The concentrations of MgCl$_2$ are indicated at the top of each lane. The markers are *Hae*III-digested φX174-RF (250 ng).

depending on the sequence being amplified and the nature of the primers. In most cases, best results with genomic targets are obtained in reactions with 0.5–1.0 mM free magnesium. (Because deoxynucleotide triphosphates quantitatively bind Mg^{++}, the amount of free Mg^{++} available for *Taq* polymerase is the concentration of MgCl$_2$ less the total concentration of dNTPs.) The effect of magnesium concentration on product specificity is shown in Figure 1 where there is a clear Mg^{++} optimum.

Deoxyribonucleotide triphosphates do not seem to affect specificity, but preliminary evidence suggests that lower concentrations may substantially improve the fidelity of *Taq* polymerase. Of each dNTP, 200 μM is sufficient substrate to synthesize over 20 μg of DNA in a 100 μl reaction and generally offers a good compromise between yield and fidelity.

The optimal concentrations for PCR primers vary widely depending on the sequences of the primers, their intended target, sequence complexity of the sample DNA, and the amount of target DNA initially present in the sample. For most genomic targets, primer concentrations in the range of 0.1–1.0 μM each seem to give the best results. The consequence of excessive primer levels is not only loss of specificity, but also an increased probability of "primer dimer" formation. Primer dimer is often seen as an intense band with a length that is the sum of the two primers. It is apparently the result of a rare event where the 3' ends of the primers come in close proximity and the polymerase extends one of the primers over the other. This is an ideal PCR template and, because of its short length, can rapidly become the predominant product in the reaction.

The concentration of enzyme typically used in PCR is 2 units/100 μl of reaction volume. For amplifications involving DNA samples with high sequence complexity, there is an optimal concentration of *Taq* polymerase, usually 1–4 units/100 μl of reaction volume. Increasing the amount of enzyme beyond this level usually results in nonspecific PCR products and a reduced yield of the target fragment (Saiki et al. 1988).

Cycling Parameters

PCR is performed by incubating the samples at three temperatures corresponding to the three steps in a cycle of amplification: denaturation, annealing, and extension. Typically, the double-stranded DNA is denatured by briefly heating the sample to 90–95ºC, the primers are allowed to anneal to their complementary sequences by briefly cooling to 40–60ºC, followed by heating to 72ºC to extend the annealed primers with *Taq* polymerase. The time of incubation at 72ºC will depend on the length of target being amplified. In most cases, the ramp time, or the time taken to change from one temperature to another, is not important, and the fastest ramps attainable are used to minimize the cycle time.

Insufficient heating during the denaturation step is one of the most common causes of failure in a PCR reaction. It is important that the temperature of the reaction reach a sufficiently high temperature for strand separation to occur. For most PCR fragments, this will be in the range of 90–95ºC; recently several laboratories have reported successful amplifications with 80–85ºC denaturation. Of course, this temperature will depend on the length and GC content of the fragment. As soon as the sample reaches the denaturation temperature, it can be cooled to the annealing temperature. Extensive denaturation is unnecessary, and limited exposure to elevated temperatures helps maintain maximum polymerase activity throughout the reaction.

The temperature at which annealing is done depends on the length and GC content of the primers. A temperature of 50–60ºC is a good starting point for typical 20-base oligonucleotide primers with about 50% GC content. Because of the very large molar excess of primers present in the reaction mix, hybridization is almost instantaneous, and long incubation at the annealing temperature is not required. Extensive incubation can lead to the production of nonspecific products. It is often possible to anneal the primers at 72ºC, the temperature at which primer extension occurs. In addition to simplifying the procedure to a two-temperature cycle, annealing at 72ºC may further improve the specificity of the reaction.

Primer extension at 72ºC is very close to the temperature of maximum activity for the *Taq* DNA polymerase. The incubation time at that temperature depends on the length of the DNA segment being amplified. Allowing 1 minute for every 1000 bp of target is usually adequate, and even shorter incubation times can be tried. The primer extension step can be eliminated altogether if the target sequence is 150 bases or less. During the thermal transition from annealing to denaturation, the sample will be within the 70–75ºC range for the few seconds required to extend the annealed primers completely. As in the annealing step, excessive incubation will usually lead to the production of nonspecific amplification products (Saiki et al. 1988).

A temperature profile that usually works well is to denature at 93ºC for 15–30 seconds, anneal at 55ºC for 15–30 seconds, and extend at 72ºC for 30–60 seconds. This profile can be run as a "step-cycle" program on a Perkin-Elmer Cetus Thermal Cycler with a single cycle time of 3–4 minutes. When used with the standard buffer described above, it is capable of amplifying a wide range of products from genomic DNA with excellent specificity (Fig. 2).

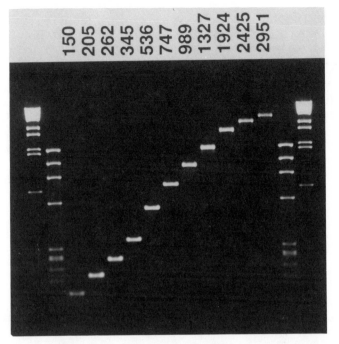

Figure 2 Amplification of β-globin fragments ranging from 150 to 2951 bp. Samples of 100 μl containing standard buffer, 200 μM each dNTP, and 250 nM each primer, 100 ng of human genomic DNA, and 2.5 units of *Taq* polymerase were subjected to 30 cycles of amplification. Of each sample, 2 μl was resolved on a 1.6% agarose gel and visualized by ethidium bromide fluorescence. The lengths of the products are indicated at the top of each lane. The markers are *Bst*EII-digested bacteriophage λ (500 ng) and *Hae*III-digested φX174-RF (250 ng).

SUMMARY

Because of the complex interactions among the components of a PCR (particularly between the primers and the DNA sample) and a wide variety of applications in which this technique is being used, it is probably impossible to specify one set of reaction conditions that will be optimal in all situations. Nevertheless, the general conditions described here have usually provided satisfactory results. When necessary, however, minor adjustments to these parameters will often transform a marginal PCR reaction into one with excellent specificity and yield.

ACKNOWLEDGMENTS

I would like to thank the members of the Cetus PCR Research Division, particularly D. Gelfand and H. Erlich, for their continuing advice and assistance.

REFERENCES

Innis, M.A., K.B. Myambo, D.H. Gelfand, and M.A.D. Brow. 1988. DNA sequencing with *Thermus aquaticus* DNA polymerase and direct sequencing of polymerase chain reaction-amplified DNA. *Proc. Natl. Acad. Sci.* **85:** 9436.

Saiki, R.K., D.H. Gelfand, S. Stoffel, S.J. Scharf, R. Higuchi, G.T. Horn, K.B. Mullis, and H.A. Erlich. 1988. Primer-directed enzymatic amplification of DNA with a thermostable DNA polymerase. *Science* **239:** 487.

Detection of Chromosomal Translocations and Minimal Residual Disease by Polymerase Chain Reaction

M.-S. Lee,[1] K.-S. Chang,[1] F. Cabanillas,[2] E.J. Freireich,[3] J.M. Trujillo,[1] and S.A. Stass[1]

[1]Hematopathology Program, Division of Laboratory Medicine, [2]Department of Hematology, and [3]Adult Leukemia Research Program, Division of Medicine, University of Texas M.D. Anderson Cancer Center, Houston, Texas 77030

Reciprocal chromosomal translocations are frequently observed in hematopoietic neoplasia, such as the t(14;18) translocation in follicular lymphoma (FL) and the Philadelphia (Ph') chromosome in chronic myelogenous leukemia (CML) (Rowley 1973; Yunis et al. 1982). Taking advantage of the unique molecular characteristics of the t(14;18) and the t(9;22) translocations, we and other investigators have utilized polymerase chain reaction (PCR) to amplify the hybrid DNA sequences of the t(14;18) and the unique mRNA sequences specific for the Ph'-positive leukemias (Lee et al. 1987, 1988; Crescenzi et al. 1988; Kawasaki et al. 1988). The PCR technique has made it possible to detect extremely small numbers of cells carrying chromosomal translocations. This technique opens a new scope of investigation to explore minimal residual disease and the mechanism of disease recurrence.

DNA Sequence Amplification of the t(14;18) Translocation

The t(14;18) translocation occurs in approximately 90% of FL and 20% of diffuse large-cell lymphomas (DLCL). In 70% of patients with FL, the breakpoints on chromosome 18 are clustered within a 150-bp major breakpoint clustering region (mbr) (Tsujimoto et al. 1985). The breakpoints on chromosome 14 always occur at the 5' end of one of the J segments (Tsujimoto et al. 1985). Since the breakpoints on both chromosomes are limited to a small DNA fragment, we were able to devise two universal primers for PCR amplification. The unique consequence of applying the PCR technique to a

31

chromosomal translocation is that it preferentially amplifies the hybrid sequences of the translocation; thus, it permits the detection of one abnormal cell among several hundred thousand normal cells. Utilizing this highly sensitive assay, we could detect minimal numbers of circulating cells carrying the t(14;18) in early stages of FL and in clinical remission (Fig. 1).

To study the mechanism of frequent and late relapse observed in low-grade FL (follicular small cleaved cell lymphoma and follicular mixed cell lymphoma), we had analyzed the blood samples obtained from 21 randomly selected patients who had achieved continuous remission for various periods of time. Evidence of minimal residual cells with the t(14;18) were demonstrated by our PCR assay in nine of ten in remission less than 2 years, in four of nine in remission for 2–5 years, and in one of two in remission for more than 5 years.

We then extended our investigation to intermediate-grade lymphomas: follicular large-cell lymphoma (FLCL) and DLCL. We had analyzed the remission blood samples from three patients with FLCL. Three of three were positive by our PCR

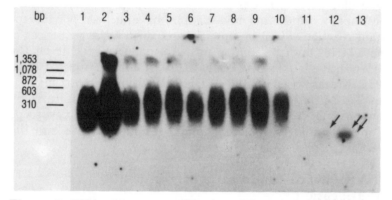

Figure 1 DNA sequence amplification of the t(14;18) translocation. Circulating cells carrying the t(14;18) are detected in the pretreatment blood samples (lanes *1–5*) and in the remission blood samples (lanes *6–10, 12,* and *13*). (Lane *1*) FSCL, stage IAE; (lane *2*) FSCL, stage IIIA; (lane *3*) FMxL, stage IIA; (lane *4*) FSCL, stage IIA; (lane *5*) FMxL, stage IIIA; (lane *6*) FSCL, remission for 2 years; (lane *7*) FSCL, remission for 6.5 years; (lane *8*) FSCL, remission for 2 years; (lane *9*) FLCL, remission for 2 years; (lane *10*) FLCL, remission for 5 years; (lane *11*) normal control; (lane *12*) FMxL, remission for 2 years; and (lane *13*) FSCL, remission for 3 years (FSCL: follicular small cleaved cell lymphoma; FMxL: follicular mixed cell lymphoma; FLCL: follicular large-cell lymphoma).

assay. Two had been in remission for more than 5 years. We also studied three DLCL patients who had the t(14;18) breakpoint within the mbr region and had been in remission for more than 2 years. Two of three showed PCR positivity.

Our observation that minimal residual cells carrying the t(14;18) are present in patients in long-term remission raises several important questions as to whether such cells are actively proliferating, whether such cells are lymphomatous or prelymphomatous, whether the host immune mechanism plays a role in preventing disease recurrence, and whether PCR-positive patients will eventually relapse clinically.

Amplification of the Unique mRNA Specific for Ph' -positive CML

The Ph' chromosome in CML involves reciprocal translocation of the *bcr* gene and the c-*abl* gene. The breakpoints on chromosome 22 in CML are clustered within a small DNA fragment designated as the *bcr* region (Groffen et al. 1984). The breakpoints on chromosome 9 vary up to more than 200 kb (Heisterkamp et al. 1983; Bernarde et al. 1987). As a result, the fused *bcr-abl* gene shows a wide variation in its DNA configuration. It is therefore difficult to make two universal primers that would anneal to the crossover site of translocation for PCR amplification. Nevertheless, the fused *bcr/abl* gene in Ph' -positive CML consistently transcribes into two types of chimeric *bcr/abl* mRNA: the L-6 junction and the K-28 junction (Shtivelman et al. 1986). Therefore, the chimeric *bcr/abl* mRNAs are the ideal target templates for PCR amplification. However, this requires conversion of the RNA templates into DNA prior to PCR.

We had used two modified PCR techniques to detect small quantities of chimeric *bcr/abl* mRNAs in Ph' -positive CML: combination of S1 protection and PCR (Lee et al. 1988) and a combination of reverse transcription and PCR (Lee et al. 1989). We studied eight patients who had achieved complete cytogenetic remission for from 6 months to 3 years after interferon therapy. All eight patients showed evidence of minimal residual *bcr/abl* transcripts (Fig. 2), suggesting that interferon therapy suppresses the proliferation of Ph' -positive clones, but it does not eradicate the Ph' -positive stem cells completely. One of these eight patients had been off interferon therapy and remained in remission for more than 1 year. This finding raises

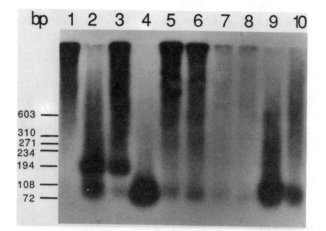

Figure 2 Reverse transcription and PCR to detect the minimal residual *bcr/abl* mRNAs in the blood samples from eight patients who had achieved complete cytogenetic remission from 6 months to 3 years after interferon therapy. (Lane *1*) Normal control; (lanes *2, 3,* and *5–10*) remission samples; (lane *4*) the blood sample in chronic phase from the same patient in lane *5*. Both the K-28 junction mRNA (155 bp) and the L-6 junction mRNA (80 bp) were detected in lanes *2* and *3*. The L-6 junction mRNA is detected in lanes *4–10*.

a question about the biological and clinical significance of the minimal residual Ph′-positive cells as detected by PCR.

SUMMARY

The PCR assay provides us with a unique opportunity to detect extremely small numbers of cells with chromosomal transloca- tions. We have observed that minimal residual cells carrying a chromosomal translocation are present even in patients in long-term remission; thus, the malignant potential of such cells has yet to be determined. Further characterization of those cells by other modalities is of ultimate importance in under- standing the biology of minimal residual disease. Long-term follow-up will eventually help determine the clinical signifi- cance of the PCR assay.

ACKNOWLEDGMENT

We wish to thank Dr. Edwin Murphy for his help in the synthesis of oligonucleotides.

REFERENCES
Bernarde, A., C.M. Rubin, C.A. Westbrook, M. Paskind, and D. Balti- more. 1987. The first intron in the human c-*abl* gene is at least 200

kilobases long and is a target for translocation in chronic myelogenous leukemia. *Mol. Cell. Biol.* **7:** 3231.

Crescenzi, M., M. Seto, G.P. Herzige, P.D. Weiss, R.C. Griffith, and S.J. Korsmeyer. 1988. Thermostable DNA polymerase chain amplification of t(14;18) chromosomal breakpoints and detection of minimal residual disease. *Proc. Natl. Acad. Sci.* **85:** 4869.

Groffen, J., J.R. Stephenson, N. Heisterkamp, A. de Klein, C.R. Bartram, and G. Grosveld. 1984. Philadelphia chromosomal breakpoints are clustered within a limited region, *bcr*, on chromosome 22. *Cell* **36:** 93.

Heisterkamp, N., J.R. Stephenson, J. Groffen, P.F. Hansen, A. de Klein, C.R. Bartram, and G. Grosveld. 1983. Localization of the c-*abl* oncogene adjacent to a translocation breakpoint in chronic myelogenous leukemia. *Nature* **306:** 239.

Kawasaki, E.S., S.S. Clark, M.Y. Coyne, S.D. Smith, R. Chaplin, O.N. Witte, and F.P. McCormick. 1988. Diagnosis of chronic myelogenous leukemia and acute leukemia by detection of leukemia-specific mRNA sequences amplified in vitro. *Proc. Natl. Acad. Sci.* **85:** 5689.

Lee, M.-S., K.-S. Chang, F. Cabanillas, E. Freireich, J.M. Trujillo, and S.A. Stass. 1987. Detection of minimal residual cells carrying the t(14;18) by DNA sequence amplification. *Science* **237:** 175.

Lee, M.-S., K.-S. Chang, H.M. Kantarjian, M. Talpaz, E. Freireich, J.M. Trujillo, and S.A. Stass. 1988. Detection of minimal residual *bcr/abl* transcripts by a modified polymerase chain reaction. *Blood* **72:** 893.

Lee, M.-S., A. LeMaistre, H.M. Kantarjian, M. Talpaz, E.J. Freireich, J.M. Trujillo, and S.A. Stass. 1989. Detection of two alternative *bcr/abl* mRNA junctions and minimal residual disease in Philadelphia chromosome positive chronic myelogenous leukemia by polymerase chain reaction. *Blood* (in press).

Rowley, J.D. 1973. A new consistent chromosomal abnormality in chronic myelogenous leukemia identified by quinacrine fluorescence and Giemsa staining. *Nature* **243:** 290.

Shtivelman, E., B. Lifshitz, R.P. Gale, B.A. Roe, and E. Canaani. 1986. Alternative splicing of RNAs transcribed from the human *abl* gene and from the *bcr-abl* fused gene. *Cell* **47:** 277.

Tsujimoto, Y., J. Gorham, J. Cossman, E. Jaffe, and C.M. Croce. 1985. The t(14;18) chromosome translocations involved in B-cell neoplasms result from mistakes in VDJ joining. *Science* **229:** 1390.

Yunis, J.J., M.M. Oken, M.I. Kaplan, K.M. Ensrud, R.R. Howe, and A. Theologides. 1982. Distinctive chromosomal abnormalities in histologic subtypes of non-Hodgkin's lymphomas. *N. Engl. J. Med.* **307:** 1231.

Application of the Polymerase Chain Reaction for the Detection of Single-base Substitutions by the RNase-A Mismatch Cleavage Method

C. Almoguera,[1] K. Forrester,[2] and M. Perucho[2]

Department of Biochemistry, State University of New York at
Stony Brook, Stony Brook, New York 11794

Advances in recombinant DNA research have led to the development of new methodologies for the diagnostic detection and characterization of a wide range of genetic alterations, including germ-line and somatic single-base substitutions, responsible for some inherited and acquired diseases (Landegren et al. 1988). We describe in this paper the properties and applications of one such technique, the ribonuclease- (RNase) A mismatch cleavage method, useful for the detection of single point mutations, with special emphasis on the diagnostic detection of mutant *ras* genes in human tumors. We also describe how the scope of this method can be greatly increased by its adaptation to in vitro gene amplification by the polymerase chain reaction (PCR).

Analysis of Point Mutations by RNase-A Mismatch Cleavage

The RNase-A mismatch cleavage method is a powerful tool for the detection and characterization of single-base substitutions in eukaryotic genes. The method uses the ability of bovine pancreatic RNase to recognize and to cleave a significant percentage of single-base mismatches in RNA:RNA (Winter et al. 1985) or DNA:RNA (Myers et al. 1985) duplexes. A homogeneously labeled RNA probe, complementary to the gene object of study, is hybridized to cellular RNA or DNA, and the hybrids are digested with RNase A. The resistant products are analyzed by electrophoresis in denaturing polyacrylamide gels and auto-

Present addresses: [1]Department of Biology, Texas A & M University, College Station, Texas 77843; [2]California Institute of Biological Research, 11099 North Torrey Pines Road, La Jolla, California 92037.

radiography. Point mutations in the gene or gene transcripts are detected by the presence of mismatch-specific subbands. Because the size of the subbands is dependent on the position of the mismatches, the mutations can be localized within the gene with an error of only a few nucleotides.

The use of long RNA probes allows the screening of relatively large genomic regions for the presence of previously uncharacterized mutations. In transcribed genes, the method is best applied using total cellular RNA. Mutations resulting in altered proteins can be detected by screening the coding region of the gene with a single or a few RNA probes. The detection of heterogeneous mutations in the hypoxanthine-guanine phosphoribosyltransferase (HPRT) gene of Lesch-Nyhan patients (Gibbs and Caskey 1987), of previously uncharacterized mutant ornithine transcarbamylase (Veres et al. 1987) and ornithine δ-aminotransferase (Mitchell et al. 1988) genes, and of mutant retinoblastoma (RB) genes (Dunn et al. 1988), are representative examples of this approach. Shorter RNA probes are indicated when the mutations are clustered in discrete regions of the gene, such as point mutations at the 3' border of the first noncoding exon of the c-myc gene in Burkitt's lymphomas (Cesarman et al. 1987) and at codons 12 or 61 in ras genes (see below; Winter et al. 1985).

Detection of Mutant ras Genes by the RNase-A Mismatch Cleavage Method

The RNase-A mismatch cleavage assay is useful for the detection and characterization of mutant ras genes. These highly conserved genes (c-Ha-ras, c-Ki-ras, and N-ras) encode GTP-binding proteins possessing intrinsic GTPase activity, which presumably are involved in the transduction of growth or differentiation signals through the cellular membrane. ras genes, activated by somatic point mutations resulting in single amino acid substitutions in their encoded proteins, are often found in human (Bos 1988) and rodent (Balmain and Brown 1988) tumors. The specific association of ras mutations with the tumor phenotype, together with the potent oncogenic activity of mutated ras genes in cultured rodent cells, strongly suggest that ras mutational activation plays an important role in tumorigenesis.

Our studies on the presence and expression of ras oncogenes in human tumors by the RNase-A mismatch cleavage analysis using total cellular RNA have been reviewed recently (Perucho et al. 1989). In brief, our results indicate that somatic c-Ki-*ras* mutations are involved in a significant proportion of human carcinomas with some degree of tissue specificity and that they can be early but also late events in tumor development, with their frequency increasing during tumor progression, including the metastatic process. Because the mutations were found exclusively at codon 12 and since the method does not depend on the position of the mutations, this finding provides strong support to the concept that *ras* mutations are involved in a causative rather than in a consequential manner in tumor development and/or progression.

Other Applications of the RNase-A Mismatch Cleavage Method

In addition to single-base substitutions, the method can detect other mutations such as single- or multiple-base insertions, duplications, and deletions because they will be reflected in the RNase-A protection pattern. An extreme example is provided by large deletions in the *HPRT* or *RB* genes. In this case, the genetic alterations are detected by the absence of protected RNA bands in the gels. Other alterations resulting in quantitative changes in the gene expression levels can be detected in the same experiment, for instance, oncogene activation by gene amplification or by promoter insertion, both resulting in overexpression of the gene in the absence of structural changes. Moreover, the search for mutations in a gene provides information about its relative expression levels, regardless of the presence or absence of the mutations. Thus, we found no major differences between the steady-state levels of the c-Ki-*ras* gene in tumors versus adjacent normal tissues in all types of tumors analyzed (Forrester et al. 1987; Perucho et al. 1989), with the exception of some lung carcinomas, which showed elevated levels of c-Ki-*ras* transcripts, concomitant with the amplification of the gene (C. Almoguera et al., unpubl.).

The RNase-A mismatch method also can provide information on the relative expression levels of both normal and mutant alleles in the same cell or tissue, even when they differ by a single nucleotide. This is accomplished by comparison of the intensities between the protected and cleaved RNA bands, corresponding to the normal and mutant transcripts, respectively. For instance, our studies with tumor cell lines show that although the mutant c-Ki-*ras* allele is often overexpressed relative to the normal (Winter et al. 1985; Forrester et al. 1987), human carcinoma cells can be heterozygous for the mutant c-Ki-*ras* allele, because some of the cell lines analyzed (lung, pancreas, and colon) express equal levels of mutant and normal transcripts (Perucho et al. 1989 and in prep.).

The technique has also found applications for the study of a variety of problems in the molecular genetics of RNA viruses. Besides the detection of single point mutations with important phenotypic consequences, such as those associated with monoclonal-antibody-resistant variants of influenza virus (Lopez-Galindez et al. 1988), and the analysis of intertypic recombination in poliovirus (Kirkegaard and Baltimore 1986), the RNase-A mismatch cleavage method is also useful for the study of genetic variability in RNA viruses. We have used the hemagglutinin gene of influenza virus as a model system (Lopez-Galindez et al. 1988). A RNA probe corresponding to a reference strain is hybridized to RNA prepared from virion particles or from cells infected by other strains. Digestion of the RNA hybrids with RNase A generates a pattern of mismatch-specific RNA subbands that is characteristic of each viral strain. Comparison of these "fingerprints" permits the cataloging and subtyping of viral isolates. The approach provides an amount of information intermediate between analysis of restriction endonuclease polymorphisms by Southern blot analysis and nucleotide sequencing and offers the great advantage of its simplicity. Comparative analysis of the RNase-A mismatch patterns between different viral strains is therefore useful for studies of the epidemiology and evolution of RNA viruses.

Analysis of Point Mutations by the RNase-A Mismatch Cleavage Method in Combination with PCR

Although the RNase-A mismatch cleavage method using total cellular RNA offers a series of advantages described above, it also suffers from some disadvantages. The main disadvantage is the need for sufficient amounts of tissue for RNA preparation. In addition, since frozen tumor specimens are used for oncogene detection, there is an uncertainty in the nature of the tissue analyzed because of the presence of variable amounts of normal tissue. Finally, RNA preparation is more laborious and requires a better tissue quality than that required for preparation of DNA. Although the RNase-A mismatch cleavage method was described utilizing total genomic DNA (Myers et al. 1985), for technical reasons, its main application was restricted to the analysis of cloned DNA sequences (Myers et al. 1985; Winter et al. 1985; Chebab et al. 1986). However, the advent of in vitro gene amplification by PCR (Saiki et al. 1985; Mullis and Faloona 1987) provided an alternative to circumvent this problem. Specific sequences present in total genomic DNA can be easily amplified to amounts equivalent to cloned gene fragments, facilitating the direct application of the RNase-A mismatch cleavage method because of the increased concentration of target sequences. In addition, the PCR requires very small amounts of DNA and therefore of tissue. Moreover, the possibility of using formalin-fixed, paraffin-embedded tissue to purify DNA suitable for some recombinant DNA procedures (Goelz et al. 1985), including PCR (Impraim et al. 1987), also widened the scope of the approach.

In collaboration with D. Shibata and N. Arnheim, we adapted the RNase-A mismatch cleavage analysis to the study of c-Ki-*ras* mutations present in formalin-fixed, paraffin-embedded tumor specimens after amplification of c-Ki-*ras* sequences by PCR. The method was simplified further by the observation that the tissue contained in a single 5–10-μm section of the paraffin blocks yielded sufficient DNA for PCR, without the need of purification by biochemical means (Shibata et al. 1988). The uncertainty due to the contaminant normal tissue present in freshly frozen tumor specimens is thus eliminated, because the nature of the tissue analyzed can be accurately determined by staining adjacent sections of the paraffin blocks.

The approach works as follows: a 5–10-μm section is cut from the paraffin block and stained with hematoxylin and eosin for histopathological characterization. Contiguous sections are placed in microtubes, deparaffinized by treatment with xylene, and the tissue is desiccated after removing the xylene by two ethanol washes. The dried sample is boiled in distilled water to release the DNA from the formalin-fixed tissue and then subjected to successive cycles of amplification by the PCR. The oligonucleotide primers are designed to amplify the DNA sequences spanning the first coding exon of the c-Ki-*ras* gene. The use of amplimers outside the coding exon avoids the potential problem of coamplifying the c-Ki-*ras* pseudogene and facilitates the analysis because the RNA probe is included within the amplified region. The amplified DNA is then hybridized to the RNA probe, digested with RNase A, and the resistant products are analyzed by denaturing polyacrylamide gel electrophoresis as before. Mutations at codon 12 of the c-Ki-*ras* first coding exon result in the generation of RNA subbands by RNase-A cleavage at the single-base mismatches in the RNA:amplified DNA hybrids. The procedure is schematized in Figure 1.

Using this approach, we have described the presence of c-Ki-*ras* genes with codon 12 mutations in the majority (21 of 22) of human pancreatic carcinomas, most of which were obtained as paraffin-embedded samples collected from surgical resections and autopsies (Almoguera et al. 1988). These findings have been confirmed by J. Bos and colleagues (Smit et al. 1988),

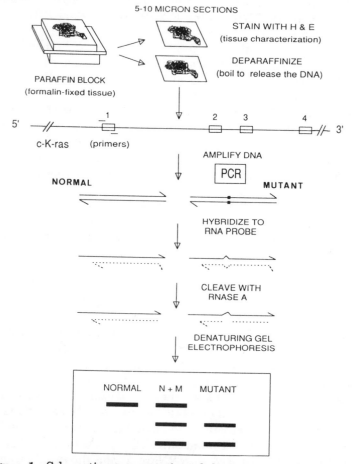

Figure 1 Schematic representation of the method to detect single point mutations in genes from formalin-fixed, paraffin-embedded tissue by PCR and RNase-A mismatch cleavage.

using a similar procedure (PCR from paraffin blocks and mutation detection by oligonucleotide hybridization). We have extended these studies and have shown that the procedure can be applied for the diagnostic detection of mutant c-Ki-*ras* genes in fine-needle aspirates of pancreatic masses, which often are the earliest tissue available for diagnosis of pancreatic carcinoma (D. Shibata et al., in prep.). The procedure also facilitates oncogene detection in multiple tissues from the same cancer patients (i.e., primary and metastatic tumors), which are often available in paraffin blocks prepared from necropsies (Almoguera et al. 1988). The method should facilitate further studies on the role of somatic *ras* mutations in neoplastic development and tumor progression and could have useful applications for cancer prognosis.

CONCLUSIONS

We have reviewed the properties and applications of the RNase-A mismatch cleavage method to detect and characterize a variety of genetic alterations, including single-base substitutions. We have discussed the advantages of using total cellular RNA and the increase in the applicability and scope of the technique by using PCR-amplified DNA. However, the method suffers from two major disadvantages: First, only a portion of all single-base mismatches are cleaved by the enzyme, and second, the extent of cleavage varies depending on the mismatch (Myers et al. 1985; Winter et al. 1985; Lopez-Galindez et al. 1988). As a result, only about 50% of single-base substitutions are detectable, and only when the mismatch is efficently cleaved is it possible to estimate the relative expression levels of normal and mutant alleles. In some situations, it is possible to circumvent these limitations. For instance, the extent of mismatch cleavage can be increased by using a probe that generates a G:U mismatch contiguous to the non- (or poorly) cleaved mismatch (Forrester et al. 1987). Moreover, although RNases other than RNase A appear unable to cleave single-base mismatches (Myers et al. 1985; M. Perucho et al., unpubl.), the proportion of cleavable mismatches could be, in principle, increased by using mixtures of RNase A from other species, which exhibit different sequence specificities for cleavage or different affinities for double-stranded RNA (Beintema 1987).

Despite these limitations, the technique has proven useful for the study of a variety of problems in the molecular genetics of RNA viruses, hereditary diseases, and malignancy. Thus, the approach that we have developed to detect single point mutations in the c-Ki-*ras* gene from formalin-fixed, paraffin-embedded samples by a combination of in vitro gene amplification by PCR and the RNase-A mismatch cleavage analysis has direct applications for cancer diagnosis, including early cancer detection. The possibility of detecting genetic alterations as small as single-base substitutions, using archival tissue samples that can be stored at room temperature for many years, should have wide applications for molecular pathology.

ACKNOWLEDGMENTS

This work was supported by National Institutes of Health grants CA-33021 and CA-38579 awarded by the National Cancer Institute and by a grant from Toyo Jozo, Ltd.

REFERENCES

Almoguera, C., D. Shibata, K. Forrester, J. Martin, N. Arnheim, and M. Perucho. 1988. Most human carcinomas of the exocrine pancreas contain mutant c-K-*ras* genes. *Cell* **53:** 549.

Balmain, A. and K. Brown. 1988. Oncogene activation in chemical carcinogenesis. *Adv. Cancer Res.* **51:** 147.

Beintema, J.J. 1987. Structure, properties and molecular evolution of pancreatic-type ribonucleases. *Life Chem. Rep.* **4:** 333.

Bos, J.L. 1988. The *ras* gene family and human carcinogenesis. *Mutat. Res.* **195:** 255.

Cesarman, E., R. Dalla-Favera, D. Bentley, and M. Groudine. 1987. Mutations in the first exon are associated with altered transcription of c-*myc* in Burkitt lymphoma. *Science* **238:** 1272.

Chebab, F.F., G.R. Honig, and Y.W. Kan. 1986. Spontaneous mutation in β-thalassaemia producing the same nucleotide substitutions as that in a common hereditary form. *Lancet* **I:** 3.

Dunn, J.M., R.A. Phillips, A.J. Becker, and B.L. Gallie. 1988. Identification of germline and somatic mutations affecting the retinoblastoma gene. *Science* **241:** 1797.

Forrester, K., C. Almoguera, K. Han, W.E. Grizzle, and M. Perucho. 1987. Detection of high incidence of K-*ras* oncogenes during human colon tumorigenesis. *Nature* **327:** 298.

Gibbs, R.A. and C.T. Caskey. 1987. Identification and localization of mutations at the Lesch-Nyhan locus by ribonuclease A cleavage. *Science* **236:** 303.

Goelz, S.E., S.R. Hamilton, and B. Vogelstein. 1985. Purification of DNA from formaldehyde fixed and paraffin-embedded human tissue. *Biochem. Biophys. Res. Commun.* **130:** 118.

Impraim, C.C., R.K. Saiki, H.A. Erlich, and R.L. Teplitz. 1987. Analysis of DNA extracted from formalin-fixed, paraffin-embedded tissues by enzymatic amplification and hybridization with sequence-specific oligonucleotides. *Biochem. Biophys. Res. Commun.* **142:** 710.

Kirkegaard, K. and D. Baltimore. 1986. The mechanism of RNA recombination in poliovirus. *Cell* **47**: 433.

Landegren, U., R. Kaiser, C.T. Caskey, and L. Hood. 1988. DNA diagnostics — Molecular techniques and automation. *Science* **242**: 229.

Lopez-Galindez, C., J.A. Lopez, J.A. Melero, L. De la Fuente, C. Martinez, J. Ortin, and M. Perucho. 1988. Analysis of genetic variability and mapping of point mutations in influenza virus by the RNAse A mismatch cleavage method. *Proc. Natl. Acad. Sci.* **85**: 3522.

Mitchell, G.A., L.C. Brody, J. Looney, G. Steel, M. Suchanek, C. Dowling, V. Der Kaloustian, M. Kaiser-Kupfer, and D. Valle. 1988. An initiator codon mutation in the ornithine-δ-aminotransferase causing gyrate atrophy of the choroid and retina. *J. Clin. Invest.* **81**: 630.

Mullis, K. and F. Faloona. 1987. Specific synthesis of DNA *in vitro* via a polymerase catalysed reaction. *Methods Enzymol.* **155**: 335.

Myers, R.M., Z. Larin, and T. Maniatis. 1985. Detection of single base substitutions by ribonuclease cleavage at mismatches in RNA:DNA duplexes. *Science* **230**: 1242.

Perucho, M., K. Forrester, C. Almoguera, S. Kahn, C. Lama, D. Shibata, N. Arnheim, and W.E. Grizzle. 1989. Expression and mutational activation of the c-Ki-*ras* gene in human carcinomas. *Cancer Cells* **7**: 137.

Saiki, R., S. Sharf, F. Faloona, K. Mullis, G. Horn, H.A. Erlich, and N. Arnheim. 1985. Enzymatic amplification of β-globin genomic sequences and restriction site analysis for diagnosis of sickle cell anemia. *Science* **230**: 1350.

Shibata, D., N. Arnheim, and J. Martin. 1988. Detection of human papilloma virus in paraffin-embedded tissue using the polymerase chain reaction. *J. Exp. Med.* **167**: 225.

Smit, V.T., A.J. Boot, A.A. Smits, G.J. Fleuren, C.J. Cornelisse, and J.L. Bos. 1988. K-*ras* codon 12 mutations occur very frequently in pancreatic adenocarcinomas. *Nucleic Acids Res.* **16**: 7773.

Veres, G., R.A. Gibbs, S.E. Scherer, and C.T. Caskey. 1987. The molecular basis of the sparse fur mouse mutation. *Science* **237**: 415.

Winter, E., F. Yamamoto, C. Almoguera, and M. Perucho. 1985. A method to detect and characterize point mutations in transcribed genes: Amplification and overexpression of the mutant c-K-*ras* allele in human tumor cells. *Proc. Natl. Acad. Sci.* **82**: 7575.

Diagnosis of Monogenic Disease Based on Polymerase Chain Reactions

H.H. Kazazian, Jr.

Genetics Unit, Department of Pediatrics, The Johns Hopkins Hospital
Baltimore, Maryland 21205

Since the fall of 1987, polymerase chain reaction (PCR) technology has had a revolutionary impact on the prenatal diagnosis of single-gene disorders and on carrier testing for these disorders. PCR technology has not yet expanded the repertoire of diseases that can be detected, but it has greatly expanded the options of the laboratory diagnostician. At Johns Hopkins Hospital, PCR has allowed us to diagnose disorders with greater speed, greater accuracy, and with greater technical flexibility. Examples illustrating each of these improvements follow. Before October 1987, we carried out prenatal diagnosis of sickle cell anemia by Southern blot analysis for the mutation, which usually required 2 or more weeks from the date of fetal sampling. Since October 1987, we have carried out these diagnoses by PCR techniques in 2–4 days after fetal sampling. Prior to October 1987, nearly all of our prenatal diagnoses of β-thalassemia were accomplished through indirect detection via linked DNA polymorphisms in the β-globin cluster. Again, this work usually took 2–4 weeks to accomplish. Since October 1987, all our prenatal diagnoses of β-thalassemia have been carried out by direct detection of the disease-producing mutations after PCR amplification of regions of the β-globin gene. These methods provide increased accuracy and faster diagnosis, usually in 1 week. Improved technical flexibility has meant that (1) if we are unable to determine a β-thalassemia mutation in one parent or another, we can study DNA polymorphisms in the β-globin cluster in 1 or 2 days or (2) we can determine the extent of maternal contamination quickly using sequence differences between mother and fetus. These are some examples of what PCR has meant to the gene diagnostic enterprise.

Use of PCR in Direct Detection of Point Mutations
Three techniques that are used following PCR in the diagnosis of point mutations, e.g., dot-blot hybridization (Saiki et al.

47

1987), restriction analysis (Chehab et al. 1987; Kogan et al. 1987), and direct sequencing (Wong et al. 1987; Engelke et al. 1988; Saiki et al. 1988), can best be discussed in the context of their use in the diagnosis of β-globin disorders. Diagnosis of sickle cell anemia is carried out at Johns Hopkins by (1) boiling fetal cells obtained by chorion villus sampling or amniocentesis in 2 M NaCl and 0.1 N NaOH, (2) PCR amplification of a 725-bp region at the 5' end of the β-globin gene using 30 cycles of PCR and 120-second chain extension time at 72°C, (3) digestion of a 725-bp amplification product with CvnI, (4) Nu-sieve agarose electrophoresis of the digestion product, (5) ethidium bromide staining of the DNA fragments, and (6) detection of DNA fragments under UV light and photography of the band pattern (Fig. 1). The primers used in the reaction were chosen such that the amplification product contains two constant CvnI sites. Thus, at least three bands are visualized in every digested sample. Each sickle β-globin gene has a 381-bp signature, whereas the $β^A$-globin gene is demonstrated by the cleaved products of this fragment and the 201-bp and 180-bp fragment. Both $β^S$ and $β^A$ genes always demonstrate 256-bp and 88-bp fragments. Because of the clarity and high information content of the results, we prefer restriction analysis of the PCR product to dot-blot hybridization using oligonucleotides specific for the $β^S$ mutation. However, a major limitation (also a limitation of Southern blot analysis) is the failure of the method to detect a deletion of the β-globin gene that includes the sixth codon, the site of the $β^S$ mutation. An individual heterozygous for such a deletion and for a $β^S$-globin gene has the same band pattern as an individual with sickle cell anemia (two $β^S$-globin genes) in this analysis. In a patient in whom the diagnosis is in doubt, the true diagnosis can best be obtained by studying the patient's parents. If both parents prove to be $β^Aβ^S$ by this technique, the diagnosis of the patient is sickle cell anemia. If one parent is $β^A$ only while the other parent is $β^Aβ^S$, the diagnosis of the patient is heterozygosity for the $β^S$ gene and a deletion involving the β-globin gene.

Direct detection of β-thalassemia mutations became theoretically possible after the mutations leading to the disease were characterized (Kazazian and Boehm 1988). Key ethnic groups at risk for β-thalassemia in their offspring are Mediterraneans, Middle Easterners, Asian Indians, Chinese, and Blacks. As of late 1988, over 60 β-thalassemia mutations are known (Gonzalez-Redondo et al. 1988; Kazazian and Boehm 1988), and the al-

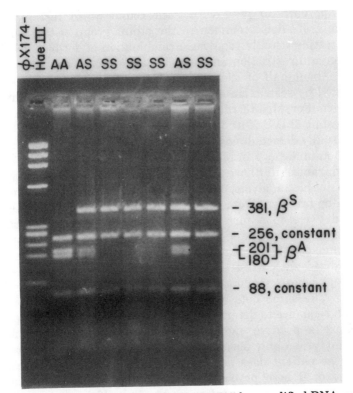

Figure 1 *Cvn*I map of a β-globin region 725-bp-amplified DNA product that contains the sequence altered by the sickle cell mutation. The 5′ and 3′ endpoints of amplified product are determined by the two primers used in the amplification reaction. Arrows represent *Cvn*I sites, and the arrow with an asterisk is the *Cvn*I site destroyed by the β^S mutation.

leles leading to well over 99% of the β-thalassemia genes in the world are characterized. Remaining alleles will be rare or present in ethnic groups with small populations at risk. Because each affected ethnic group has its own battery of β-thalassemia mutations and, in general, four to six alleles make up greater than 90% of β-thalassemia genes in any particular ethnic group (Kazazian and Boehm 1988), the job of detecting the disease-producing mutations in both members of an at-risk couple is simplified.

Often a childless couple is referred for testing because both members appear to have the β-thalassemia trait in screening tests of their red cell volume and hemoglobin A$_2$ concentration.

49

The approach to take with each couple is to test for the presence of alleles common in the ethnic group of the couple. This testing usually requires a combination of dot-blot hybridization and restriction analysis. Rarely, genomic sequencing is used. About half of the over 60 β-thalassemia alleles can be detected by restriction endonuclease analysis of the amplified product. For Mediterranean couples, the nonsense codon 39, frameshift 6; IVS-2, nt 1; IVS-2, nt 745; and IVS-1, nt 6 alleles are usually detected most easily by this method. The remaining common mutations in Mediterraneans (IVS-1, nt 110; IVS-1, nt 1; and frameshift 8) are all detected by dot-blot analysis.

In dot-blot analysis, 5–10% of the amplified product (usually a 725-bp fragment from the 5' one-half of the β-globin gene) is dotted twice on a nitrocellulose filter. Two dots are usually tested from each individual for control purposes, to visualize the same positive or negative result with each dot. The dots are hybridized with a ^{32}P-end-labeled oligonucleotide (19-mer or 20-mer) whose sequence is specific for the mutation under study. The same dots are then hybridized with a normal β-globin sequence to determine that sufficient DNA was dotted to produce a positive signal. If the assay is negative with a mutant probe and positive with the normal probe, we know that the patient does not carry the particular mutation being analyzed. If both probes give positive results, the patient is heterozygous for the mutation under study. If the mutant probe gives a positive result and the normal probe does not, the patient is homozygous for the analyzed mutation (see Fig. 2).

Rarely, the β-thalassemia allele is still unknown in one member of a couple after DNA of both members of the couple has been analyzed for the common alleles present in their ethnic group. At that point, key regions of the β-globin gene are sequenced to determine the unknown disease-producing mutation (Wong et al. 1987; Engelke et al. 1988; Saiki et al. 1988). Sequence analysis is carried out using T7 DNA polymerase (Sequenase) on PCR-amplified regions of genomic DNA. These regions are (1) the promoter, exon 1, intron 1, exon 2, and the 5' 90 nucleotides of intron 2 and (2) the 3' 300 nucleotides of intron 2 and exon 3. This procedure yields the unknown mutation in nearly all cases studied. However, in at least two instances, a mutation has not been found in the β-globin gene or the key 5' and 3' regions in carriers of mild β-thalassemia alleles (H.H. Kazazian, Jr. et al., unpubl.). In our hands, seven previously undescribed alleles have been found by direct se-

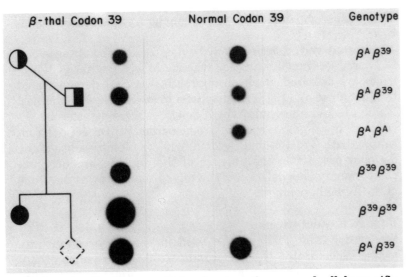

Figure 2 Dot-blot analysis showing the use of allele-specific oligonucleotide (ASO) probes on amplified β-globin DNA for prenatal diagnosis of β-thalassemia. Both parents in the pedigree at the left carry the nonsense codon 39 mutation, as demonstrated by hybridization of their amplified DNAs to the mutant ASO. Control samples, essential for monitoring the specificity of the wash, are in the third and fourth dots from the top and represent amplified DNA from individuals homozygous for normal β-globin alleles and homozygous for the nonsense codon 39 β-thalassemia allele, respectively. Amplified DNA from this couple's affected child hybridizes to mutant ASO only and from the fetus to both the mutant and normal ASOs, which demonstrates that the fetus is a carrier of β-thalassemia.

quence analysis of PCR-amplified β-globin genes (Wong et al. 1987; Kazazian and Boehm 1988). Huisman and colleagues (Gonzalez-Redondo et al. 1988) have found a comparable number of additional new alleles in populations different from those that we have studied.

Detection of Deletions by PCR
Deletions of moderate size (< 1–1.5 kb) can be detected through use of primers located 5′ and 3′ to the deletion. A 619-bp deletion in the β-globin gene is a common β-thalassemia allele among Asian Indians (Spritz and Orkin 1989). We detect this deletion using primers that yield a 1215-bp fragment in normal DNA and a 596-bp fragment when the 619-bp deletion is present. The presence of both the 1215-bp and 596-bp fragments

after PCR of genomic DNA signifies heterozygosity for the deletion.

Alternatively, deletions may be detected by the absence of a PCR product from a multiplex reaction in which other PCR products are detected. Such an approach was first demonstrated by Chehab et al. (1987) for deletions involving the $\tilde{a}$-globin loci leading to $\tilde{a}$-thalassemia. These deletions were detected by absence of the $\tilde{a}$-globin specific product from PCR reactions containing both $\tilde{a}$-globin and β-globin primers. The presence of the β-globin product in the absence of the $\tilde{a}$-globin product indicates homozygous deletion of the region containing at least one of the $\tilde{a}$-globin primers.

Future Expectations

We predict that PCR will be used in carrier screening programs. Specifically, the best candidate diseases will be those with high incidence in one or more population groups, significant morbidity for affected individuals, and those for which there is knowledge about the gene of interest and the mutations that account for essentially all of the mutant alleles. Carrier screening for β-thalassemia and sickle cell anemia by DNA methods is possible, but non-DNA tests are still cheaper and simpler than any gene diagnosis test.

Carrier screening by PCR for Tay-Sachs alleles in the adult Ashkenazi Jewish population has strong possibilities. Recent data suggest that two alleles producing defects in the $\tilde{a}$-chain gene of hexosaminidase A may account for essentially all Tay-Sachs genes in the Ashkenazi Jewish population (Arpaia et al. 1988; Myerowitz 1988; Myerowitz and Costigan 1989). If this is so, then screening for carriers of infantile Tay-Sachs disease genes might best be carried out by PCR analysis.

The alleles accounting for 60% of phenylketonuria (PKU) genes in the northern European population have been characterized (DiLella et al. 1986, 1987). No other practical means for detecting carriers of PKU other than gene analysis exists. Thus, it is probable that some individuals at a higher risk of being carriers because of their family histories will desire carrier screening when the alleles producing PKU are completely characterized.

When the cystic fibrosis (CF) gene is discovered and alleles producing CF are characterized, this disorder will be a prime candidate for carrier screening by PCR techniques. Carrier screening by gene analysis will be practical if the number of alleles is small (fewer than 10) and no simple biochemical test for carrier status emerges from knowledge of the defective protein. The possibility exists that population-based screening for carriers of the CF gene using PCR could have a major influence on the incidence of this disease.

The possibilities for use of PCR in diagnosis of monogenic diseases are exceedingly great. PCR-based diagnosis by analysis of linked DNA polymorphisms is being used in diagnostic testing for hemophilia, cystic fibrosis (Feldman et al. 1988a,b), and Huntington's disease (Brock et al. 1988). It could be carried out for neurofibromatosis, retinoblastoma, adult onset polycystic kidney disease, and myotonic dystrophy, among others. PCR-based analysis of exon sequences by density gradient gel electrophoresis in various disease genes, e.g., the factor VIII:C exons in hemophilia A patients, could lead to direct diagnosis of mutant alleles in most affected individuals (Sheffield et al. 1989). New techniques in which oligonucleotides with poly(T) tails are immobilized on filters and hybridized with a biotinylated PCR fragment of the region of interest from the patient's DNA may revolutionize gene diagnosis (Saiki and Erlich 1988). This technique will simplify assay for from five to ten or more alleles in the genome of an individual as is now carried out for β-thalassemia. The technique may be used in a general genetic profile analysis, whereby the genome of an individual is tested for the presence of a number of deleterious genes, e.g., β-thalassemia alleles pertinent to his ethnic group, the sickle cell allele, the PKU alleles, the CF alleles, and the Tay-Sachs alleles. A multiplex PCR might be done to screen a number of genes simultaneously.

SUMMARY

Use of PCR for diagnosis of monogenic diseases has become well established. In 1988 at Johns Hopkins Hospital, all prenatal diagnoses for sickle cell anemia and β-thalassemia and one-half of the diagnoses for hemophilia A were carried out using PCR techniques. Overall, about one-half of the approximately 425 families studied at Johns Hopkins Hospital in 1988 received diagnostic information obtained following PCR. In 1989 at Johns Hopkins, one-half of all Duchenne's muscular dystrophy diagnoses, one-half or more of CF diagnoses, and one-half or more of Tay-Sachs screening will be carried out using PCR techniques. The use of Southern blot analysis in this diagnostic work will not be eliminated, but we expect that the use of Southern blots in 1989 will decline to about 30–40% of its use in 1987. PCR techniques will probably replace about 90% of Southern blot analysis in the diagnoses of monogenic diseases within 3–5 years. However, care will have to be taken to maintain the highest diagnostic accuracy during this transition period.

REFERENCES

Arpaia, E., A. Dumbrille-Ross, T. Maler, K. Neote, M. Tropak, C. Troxel, J.L. Stirling, J.S. Pitts, B. Bapat, A.M. Lamhonwah, D.J. Mahuran, S.M. Schuster, J.T.R. Clarke, J.A. Lowden, and R.A. Gravel. 1988. Identification of an altered splice site in Ashkenazi Tay-Sachs disease. *Nature* 333: 85.

Brock, D.J.H., I. McIntosh, A. Curtis, and F.A. Millan. 1988. Use of the polymerase chain reaction in prenatal exclusion testing for Huntington's disease. *Am. J. Hum. Genet.* 43: A79. (Abstr.)

Chehab, F., M. Dogerty, S. Cai, Y.W. Kan, S. Cooper, and E. Rubin. 1987. Detection of sickle cell anemia and thalassemia. *Nature* 329: 293.

DiLella, A.G., J. Marvit, K. Brayton, and S.L.C. Woo. 1987. An amino-acid substitution involved in phenylketonuria is in linkage disequilibrium with DNA haplotype 2. *Nature* 327: 333.

DiLella, A.G., J. Marvit, A.S. Lidsky, F. Guttler, and S.L.C. Woo. 1986. Tight linkage between a splicing mutation and a specific DNA haplotype in phenylketonuria. *Nature* 322: 799.

Engelke, D.R., P.A. Hoener, and F.S. Collins. 1988. Direct sequencing of enzymatically amplified human genomic DNA. *Proc. Natl. Acad. Sci.* 85: 544.

Feldman, G.L., R. Williamson, A.L. Beaudet, and W.E. O'Brien. 1988a. Prenatal diagnosis of cystic fibrosis by DNA amplification for detection of KM-19 polymorphism. *Lancet* II: 102.

Feldman, G.L., W.E. O'Brien, B. Durtschi, P. Gardner, R. Williamson, and A.L. Beaudet. 1988b. Prenatal diagnosis of cystic fibrosis (CF) using the polymerase chain reaction (PCR) for detection of the KM-19 polymorphism. *Am. J. Hum. Genet.* 43: A83. (Abstr.)

Gonzalez-Redondo, J.M., T.A. Stoming, K.D. Lanclos, U.C. Gy, A. Kutlar, F. Kutlar, T. Nalcatsuji, B. Deng, I.S. Han, V.C. McKie, and T.J.H. Huisman. 1988. Clinical and genetic heterogeneity in black patients with homozygous β-thalassemia from the Southeastern United States. *Blood* 72: 1007.

Kazazian, H.H., Jr. and C.D. Boehm. 1988. Molecular basis and prenatal diagnosis of β-thalassemia. *Blood* 72: 1107.

Kogan, S.C., M. Doherty, and J. Gitschier. 1987. An improved method for prenatal diagnosis of genetic diseases by analysis of amplified DNA sequences: Application to hemophilia A. *N. Engl. J. Med.* 317: 985.

Myerowitz, R. 1988. Splice junction mutation in some Ashkenazi Jews with Tay-Sachs disease: Evidence against the single defect within this ethnic group. *Proc. Natl. Acad. Sci.* 85: 3955.

Myerowitz, R. and F.C. Costigan. 1989. The major defect in Ashkenazi

Jews with Tay-Sachs disease is an insertion in the gene for the α-chain of β-hexosaminidase. *J. Biol. Chem.* **263:** 18587.

Saiki, R.K. and H.A. Erlich. 1988. Genetic analysis of PCR amplified DNA with immobilized sequence-specific oligonucleotide probes: Reverse dot blot typing of HLA and β-globin loci. *Am. J. Hum. Genet.* **43:** A95.

Saiki, R.K., T.L. Bugawan, G.T. Horn, K.B. Mullis, and H.A. Erlich. 1987. Analysis of enzymatically amplified β-globin and HLA-DQα DNA with allele-specific oligonucleotide probes. *Nature* **324:** 163.

Saiki, R.K., D.H. Gelfand, S. Stoffel, S.J. Scharf, R. Higuchi, G.T. Horn, K.B. Mullis, and H.A. Erlich. 1988. Primer-directed enzymatic amplification of DNA with a thermostable DNA polymerase. *Science* **239:** 489.

Sheffield, V.C., D.R. Cox, S.L. Lerman, and R.M. Myers. 1989. Attachment of a CG- clamp to genomic DNA fragments by the polymerase chain reaction results in improved detection of single base changes. *Proc. Natl. Acad. Sci.* **86:** 232.

Spritz, R.A. and S.H. Orkin. 1989. Duplication by deletion accounts for the structure of an Indian deletion β-thalassemia resulting from a 1.4 kb deletion. *Nucleic Acids Res.* **10:** 8025.

Wong, C., C.E. Dowling, R.K. Saiki, R.G. Higuchi, H.A. Erlich, and H.H. Kazazian, Jr. 1987. Characterization of β-thalassemia mutations using direct genomic sequencing of amplified single copy DNA. *Nature* **330:** 384.

Identification and Screening for Phenylketonuria Mutations by Polymerase Chain Reaction

S.L.C. Woo, Y. Okano, T. Wang, and R. Eisensmith

Howard Hughes Medical Institute, Department of Cell Biology and
Institute of Molecular Genetics, Baylor College of Medicine
Houston, Texas 77030

Classic phenylketonuria (PKU) is an autosomal recessive disorder caused by an inborn error of amino acid metabolism, with an incidence of about 1 in 10,000 Caucasian births. The disease results from absence of hepatic phenylalanine hydroxylase (PAH) and causes mental retardation (for reviews, see Scriver et al. 1988; Woo 1989). We have previously reported the construction of a full-length human *PAH* cDNA clone (Kwok et al. 1985) and the use of it to identify extensive restriction-fragment-length polymorphisms (RFLPs) in the human *PAH* locus (Lidsky et al. 1985).

Characterization of Mutations Associated with Haplotypes 2 and 3 of the *PAH* gene

Twelve RFLP haplotypes at the *PAH* locus in the northern European population have been characterized, and about 90% of the PKU alleles in this population are confined to RFLP haplotypes 1–4 (Chakraborty et al. 1987). We previously established at molecular and biochemical levels that mutations in the *PAH* gene associated with haplotypes 2 and 3 result in undetectable PAH activity (DiLella et al. 1986, 1987). The mutation associated with haplotype 3 is caused by a single-base substitution at the exon 12/intron 12 boundary and comprises about 40% of mutant alleles, whereas the mutation associated with haplotype 2 is caused by a substitution of arginine for tryptophan at residue 408 in exon 12 and comprises about 20% of the mutant *PAH* genes in the northern European population.

Oligonucleotides specific for these mutations can serve as molecular probes to detect PKU carriers in the general population. The polymerase chain reaction (PCR) technique (Saiki et al. 1985; Saiki 1987) will expedite screening for mutant alleles at the *PAH* locus (DiLella et al. 1988). PCR obviates the need to analyze point mutations by conventional restriction enzyme digestion and Southern blot analysis, enabling the entire analytical procedure to be carried out in a day with submicrogram quantities of DNA.

We applied the PCR technology to the analysis of genomic DNA samples of PKU patients bearing haplotypes 2 and 3 mutant alleles (DiLella et al. 1988). The mutant haplotype 2 probe hybridized only with amplified genomic DNA from the haplotype 2 carriers and the compound heterozygotes bearing both mutant alleles (Fig. 1A). Similarly, the mutant haplotype 3 probe hybridized only with amplified genomic DNA from the haplotype 3 carriers and the compound heterozygotes (Fig. 1B). Moreover, the PCR amplification method greatly enhanced the sensitivity of the analysis, since no hybridization signal was detectable with unamplified DNA samples (Fig. 1A,B). In addition, no hybridization was detected in amplified normal genomic DNA samples with either mutant oligonucleotide probe, indicating the specificity of the analysis.

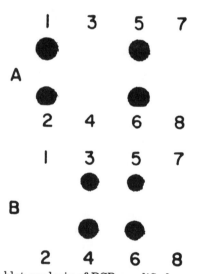

Figure 1 Dot-blot analysis of PCR-amplified genomic DNA. *A* and *B* are autoradiographs of the membranes after hybridization with the mutant haplotype 2 and 3 oligonucleotide probes, respectively. (*1* and *2*) Two mutant haplotype 2 carriers; (*3* and *4*) two mutant haplotype 3 carriers; (*5* and *6*) two haplotypes 2 and 3 compound heterozygotes; (*7* and *8*) two normal individuals.

Identification of Mutations Associated with
Haplotypes 1 and 4 of the *PAH* gene

More recently, we have identified a molecular lesion associated with haplotype 1 and 4 mutant alleles in the Swiss population, using PCR-mediated amplification techniques. Exon-containing regions of the *PAH* gene were amplified by PCR from genomic DNA of a PKU patient who is a compound heterozygote bearing a haplotype 1 and a haplotype 4 mutant allele. The amplified DNA fragments were subcloned into M13 for sequence analysis. Missense mutations were observed in exons 5 and 7, resulting in the replacement of arginine by glutamine at residues 158 and 261 of the enzyme, respectively. Direct hybridization analysis of the point mutations using specific oligonucleotide probes demonstrated that the exon 7 mutation is present in 13 of 18 haplotype 1 mutant alleles and that the exon 5 mutation is present in 2 of 6 haplotype 4 mutant alleles in the Swiss population. The mutations were not detected in normal or other haplotype mutant alleles. These results provide strong evidence for linkage disequilibrium between the mutant haplotypes and specific mutations in the *PAH* gene. Furthermore, they constitute conclusive evidence that haplotype 1 and 4 chromosomes bear multiple mutant *PAH* alleles in Caucasians.

Potential Carrier Screening for PKU in Caucasians

The specific mutant oligonucleotides were then used to perform population genetic analysis of mutant *PAH* alleles among Caucasians in various European regions. The Arg-261 to Gln-261 mutation accounts for 26% of all mutant haplotype 1 alleles in the European population, whereas the Arg-158 to Gln-158 mutation accounts for 48% of all mutant haplotype 4 alleles in this population. Since haplotype 1 and 4 mutant *PAH* alleles account for 28% and 13%, respectively, of all PKU alleles in

Table 1 Cumulative Frequency of Prevalent Mutant *PAH* Alleles in Europe

Haplotypes	Mutations	Frequencies (%)		
1	Arg–261 to Gln–261	28 x 28 =	7.8
2	Arg–408 to Gln–408	100 x 21 =	21.0
3	GT(12) to AT(12)	92 x 15 =	13.8
4	Arg–158 to Gln–158	46 x 13 =	6.0
Cumulative frequency			48.6

Europe, the current carrier detection rates of total haplotype 1 and 4 mutations are 28% x 26% = 7.2% and 13% x 48% = 6.2%, respectively. Together with the Arg-408 to Gln-408 and the intron 12 splicing mutations, the overall potential for carrier detection of PKU at the present time is 48.2% in the European population using specific oligonucleotide probes to all four mutant alleles (Table 1). It would be of great importance to characterize the mutations in the remaining prevalent mutant alleles to increase the accuracy rate to the 90% level, at which point the potential for carrier detection of PKU in the Caucasian population without a proband can be realized.

ACKNOWLEDGMENTS

This work was supported in part by National Institutes of Health grant HD-17711. S.L.C.W. is an investigator of the Howard Hughes Medical Institute.

REFERENCES

Chakraborty, R., A.S. Lidsky, S.P. Daiger, F. Güttler, S. Sullivan, A. DiLella, and S. Woo. 1987. Polymorphic DNA haplotypes at the phenylalanine hydroxylase locus and their association with phenylketonuria. *Hum. Genet.* **76**: 40.

DiLella, A.G., W.M. Huang, and S.L.C. Woo. 1988. Screening for phenylketonuria mutations by DNA amplification with the polymerase chain reaction. *Lancet* **I**: 497.

DiLella, A.G., J. Marvit, K. Brayton, and S.L.C. Woo. 1987. An amino-acid substitution involved in phenylketonuria is in linkage disequilibrium with DNA haplotype 2. *Nature* **327**: 333.

DiLella, A.G., J. Marvit, A.S. Lidsky, F. Guttler, and S.L.C. Woo. 1986. Tight linkage between a splicing mutation and a specific DNA haplotype in phenylketonuria. *Nature* **322**: 799.

Kwok, S.C.M., F.D. Ledley, A.G. DiLella, K.J.H. Robson, and S.L.C. Woo. 1985. Nucleotide sequence of a full-length complementary DNA clone and amino acid sequence of human phenylalanine hydroxylase. *Biochemistry* **24**: 556.

Lidsky, A.S., F.D. Ledley, A.G. DiLella, S.C.M. Kwok, S.P. Daiger, K.J.H. Robson, and S.L.C. Woo. 1985. Extensive restriction site polymorphism at the human phenylalanine hydroxylase locus and application in prenatal diagnosis of phenylketonuria. *Am. J. Hum. Genet.* **37**: 619.

Saiki, R.K. 1987. Genetic analysis of enzymatically amplified β globin and HLA-DQ genomic DNA with allele specific oligonucleotide probes. *Nature* **324**: 163.

Saiki, S.K., S. Scharf, F. Faloona, K. Mullis, G.T. Horn, H. Erlich, and

N. Arnheim. 1985. Enzymatic amplification of globin genomic sequences and restriction site analysis oligonucleotide probes. *Science* **230:** 1350.

Scriver, C.R., S. Kauman, and S.L.C. Woo. 1987. Mendelian hyperphenylalaninemia. *Annu. Rev. Genet.* **22:** 301.

Woo, S.L.C. 1989. Molecular basis and population genetics of phenylketonuria. *Biochemistry* **28:** 1.

Mutation Detection and Structure-Function Studies at the Ornithine δ-Aminotransferase Locus

D. Valle,[1,2] G.A. Mitchell,[3] L.C. Brody,[1,2] L.S. Martin,[2] G. Steel,[1] and I. Sipila,[4]

[1]Howard Hughes Medical Institute and the [2]Department of Pediatrics
Johns Hopkins University School of Medicine
Baltimore, Maryland 21205

[3]Division of Genetics, Saint Justine Hospital
Montreal, Canada H3T 1C5

[4]Department of Pediatrics, Children's Hospital, University of
Helsinki, Helsinki, Finland

Ornithine δ-aminotransferase (OAT, EC 2.6.1.13) is a homohexameric mitochondrial matrix enzyme that catalyzes the reversible transamination of ornithine to Δ^1-pyrroline-5-carboxylate (Valle and Simell 1983). This reaction is necessary for the catabolism of arginine and ornithine and is part of a pathway that interconnects the tricarboxylic acid and urea cycles. Deficiency of OAT in humans causes the autosomal recessive chorioretinal degeneration, gyrate atrophy (GA), of the choroid and retina. The distribution of GA is panethnic with a much higher frequency in Finns than in other populations. The clinical phenotype begins with the onset of myopia and reduced night vision in childhood. The visual fields gradually constrict, and most patients are virtually blind by the fifth to sixth decade. There is considerable interfamilial variability with some patients losing nearly all sight by the fourth decade, whereas others have functional vision into the eighth or ninth decade. Nearly all patients develop cataracts in the second to third decade. The ocular fundus shows a characteristic pattern of complete chorioretinal degeneration that begins in the periphery and gradually extends toward the posterior pole (Valle and Simell 1983). The metabolic phenotype of GA is characterized by ornithine accumulation to levels approximately tenfold that of normal in all bodily fluids. OAT activity is less than 5% of normal in cultured skin fibroblasts and a variety of other

63

cells and tissues. Heterozygotes are asymptomatic and have OAT activity ranging from 25% to 75% of normal. A few patients exhibit in vivo and/or in vitro responsiveness to pharmacologic doses of vitamin B_6 (pyridoxine).

To study the cell biology of OAT and the mutations causing GA, we have cloned and sequenced a nearly full-length human liver *OAT* cDNA and have determined the structure of the human *OAT* structural gene, which has 11 exons and spans 22 kb of DNA at 10q26 (Mitchell et al. 1988a). A cluster of nonfunctional OAT-related sequences are located on the X chromosome (Mitchell et al. 1986, 1988a; Barrett et al. 1987; Ramesh et al. 1988a). Three missense mutations of *OAT* causing GA have recently been reported. The first (found by our group and designated M1I) changes the initiation methionine to isoleucine and was found in affected members of two unrelated pedigrees of Lebanese Maronite descent (Mitchell et al. 1988b). The second was an alteration of Val-332 to methionine (V332M) in a pyridoxine-responsive patient, and a third was a conversion of Asn-54 to lysine (N54K) (Ramesh et al. 1988b). The ethnic background of the latter two patients was not provided.

Mutation Detection Strategy

We have been studying *OAT* mutations in GA probands from more than 100 pedigrees from 20 different ethnic backgrounds. The optimal approach for detection of mutations depends, in large part, on the characteristics of the gene being analyzed. Relevant features of the *OAT* gene include the following: the organization of the structural gene is known (Mitchell et al. 1988a); multiple pseudogenes (~10) are present that complicate the interpretation of genomic blots; and an expressing tissue (cultured skin fibroblasts) is available. We have made particular use of two techniques: RNase-A protection assays of patient mRNA (Mitchell et al. 1988b) and polymerase chain reaction (PCR) amplification of genomic DNA and fibroblast mRNA. RNase-A protection of patient RNA has the advantage of focusing on the coding region and has sufficient sensitivity to detect many single base-pair substitutions. However, it requires access to an expressing tissue and detects only about 50% of the point mutations. PCR amplification and sequence analysis of genomic DNA is straightforward and detects all mutations within the amplified fragments. However, it requires knowledge of gene structure and, for genes with many exons, is a large undertaking. PCR amplification of cDNA allows for scanning of several exons at once but requires access to mutant mRNA.

Initial Southern blot analysis of leukocyte DNA from probands in 40 of our GA families failed to reveal evidence of gross gene rearrangement. In Northern blots of 37 probands from this group, we identified four mRNA phenotypes: normal size and amount (78%); moderately reduced amount (14%); undetectable (6%); and abnormally sized mRNA (2%). The high percentage of mRNA-positive mutants predicts that most fall within the coding portion of the OAT message. With this information, we established an algorithm for mutation detection in *OAT*, shown in Figure 1.

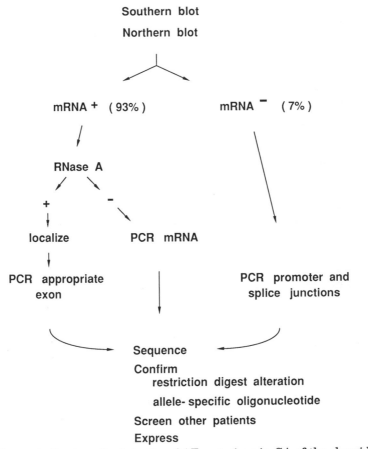

Figure 1 Strategy for detecting *OAT* mutations in GA of the choroid and retina.

We study all mRNA-positive lines by RNase-A protection. If an abnormality is detected, we amplify, clone, and sequence the predicted exons. If RNase-A protection is uninformative, or if patient fibroblasts are unavailable or are mRNA-negative, we systematically amplify, clone, and sequence OAT exons.

For genomic PCR, we employ intron oligonucleotide primers to amplify the entire exon, including the splice junctions. This avoids possible interference by OAT-related pseudogenes. To simplify the detection of genetic compounds and to avoid confusion with random *Taq* polymerase errors, we pool up to 50 PCR clones and use a combined plasmid preparation for sequence analysis. Once a particular mutation is identified, we determine its occurrence in other GA patients and normals. Finally, we construct and express the corresponding mutant *OAT* cDNA to examine the functional consequences of the mutation.

OAT Mutations

To date, we have detected nine *OAT* mutations, including seven missense mutations, one single nucleotide deletion, and one in-frame 3-nucleotide deletion that deletes Ala-184. Because GA is a prototype of the so-called Finnish genetic diseases, we were particularly interested in defining the mutation(s) responsible for GA in Finns. To our surprise, we have found that at least three mutant *OAT* alleles are segregating in Finns (Mitchell et al. 1989). One, L402P, results from a T→C transition at position 1205, accounts for about 80% of the Finn GA pedigrees in our series of 27, and, to date, has not been found in non-Finns. The second, R180T, results from a G→C transversion at nucleotide 539, is segregating in about 10% of Finn GA pedigrees, and was also found in a single American GA patient. Probands in 4 of 27 Finn GA pedigrees are genetic compounds: one for R180T/L402P and the other three for L402P and as yet undefined allele(s). Interestingly, OAT haplotype analysis with three polymorphic restriction endonuclease sites (*Eco*RI, 5′ Msp, and 3′ Msp) indicates that the R180T allele arose independently in the ancestors of the American and Finn R180T patients.

We have assessed the functional consequences of these mutations by expressing the corresponding cDNAs in Chinese hamster ovary (CHO) cells. CHO cells lack endogenous OAT activity (Valle et al. 1973) or mRNA (Mitchell et al. 1989) and, therefore, are ideal for this purpose. Wild-type human *OAT* cDNA in the expression vector of Kaufman (Wong et al. 1985) is expressed at high levels in both transient transfection assays and permanent transfectants in these cells. R180T, L402P, and A184 all produce functionally inactive OAT protein, and M1I, as predicted, blocks translation initiation.

Mutagenesis
To understand better OAT function and to develop strategies for producing an animal model of GA, we have utilized PCR with mutant oligomeric primers to construct a variety of *OAT* mutations. For example, we have mutagenized the putative pyridoxal phosphate-binding lysine to either glutamate (K292E) or arginine (K292R) and have shown by transfection of these constructs into CHO cells that this completely inactivates OAT. Cotransfection of either of the K292 mutants with wild-type OAT has inhibitory effects on the expression of the OAT activity produced by the wild-type construct (Brody et al. 1988). This result suggests that these mutant subunits may act as "suicide subunits" and could be used in transgenic animals to inhibit the expression of endogenous OAT and thereby create useful animal models of GA.

ACKNOWLEDGMENTS
A portion of the work described was supported by National Institutes of Health grant EY-02948 and Clinical Research grant RR-0052. Ms. Sandra Muscelli prepared the manuscript.

REFERENCES
Barrett, D.J., J.B. Bateman, R.S. Sparkes, T. Mohandas, I. Klisak, and G. Inana. 1987. Chromosomal localization of human ornithine aminotransferase gene sequences to 10q26 and Xp11.2. *Invest. Opthalmol. Visual Sci.* 28: 1037.
Brody, L.C., G. Mitchell, C. Wong, G. Steel, J. Looney, I. Sipila, and D. Valle. 1988. Functional consequences of synthetic and naturally occurring mutations in human ornithine-δ-aminotransferase. *Am. J. Hum. Genet.* 43: A79.
Mitchell, G.A., D. Valle, H. Willard, G. Steel, M. Suchanek, and L.C. Brody. 1986. Human ornithine-δ-aminotransferase: Cross-hybridizing fragments mapped to chromosome 10 and Xp11.1 – 21.1. *Am. J. Hum. Genet.* 39: 163A.
Mitchell, G.A., J.E. Looney, L.C. Brody, G. Steel, M. Suchanek, J. Engelhardt, H.F. Willard, and D. Valle. 1988a. Human ornithine-δ-aminotransferase: cDNA cloning and analysis of the structural gene. *J. Biol. Chem.* 263: 14228.
Mitchell, G.A., L.C. Brody, J. Looney, G. Steel, M. Suchanek, C. Dowling, V. Der Kaloustian, M.I. Kaiser-Kupfer, and D. Valle. 1988b. An initiator codon mutation in ornithine-δ-aminotransferase causing gyrate atrophy. *J. Clin. Invest.* 81: 630.

Mitchell, G.A., L.C. Brody, I. Sipila, J.E. Looney, C. Wong, J. Engelhardt, A.S. Patel, G. Steel, C. Obie, M.I. Kaiser-Kupfer, and D.
Valle. 1989. At least two mutant alleles of ornithine-δ-
aminotransferase cause gyrate atrophy of the choroid and retina in
Finns. *Proc. Natl. Acad. Sci.* **86:** 197.

Ramesh, V., R. Eddy, G.A. Burns, V.E. Shih, T.B. Shows, and J.F.
Gusella. 1988a. Localization of the ornithine aminotransferase
gene and related sequences on two human chromosomes. *Hum.
Genet.* **76:** 121.

Ramesh, V., A.E. McClatchey, N. Ramesh, L.A. Benoit, E.L. Berson,
V.E. Shih, and J.F. Gusella. 1988b. Molecular basis of ornithine
aminotransferase deficiency in B-6-responsive and -nonresponsive
forms of gyrate atrophy. *Proc. Natl. Acad. Sci.* **85:** 3777.

Valle, D. and O. Simell. 1983. The hyperornithinemias. In *The metabolic basis of inherited diseases*, 5th edition (ed. J.B. Stansbury et
al.), p. 382. McGraw-Hill, New York.

Valle, D., S.J. Downing, S.C. Harris, and J.M. Phang. 1973. Proline
biosynthesis: Multiple defects in Chinese hamster ovary cells.
Biochem. Biophys. Res. Commun. **53:** 1130.

Wong, G.G., J.S. Witek, P.A. Temple, K.M. Wilkens, A.C. Leary, D.P.
Luxenberg, S.S. Jones, E.L. Brown, R.M. Kay, E.C. Orr, C. Shoe-
maker, D.W. Golde, R.J. Kaufman, R.M. Hewick, E.A. Wang, and
S.C. Clark. 1985. Human GM-CSF: Molecular cloning of the complementary DNA and purification of the natural and recombinant
proteins. *Science* **228:** 810.

Application of Polymerase Chain Reaction to Cystic Fibrosis: Prenatal Diagnosis, Carrier Testing, and Community Genetics

R. Williamson

Department of Biochemistry and Molecular Genetics, St. Mary's
Hospital Medical School, Imperial College
London W2 1PG, United Kingdom

The mutation causing cystic fibrosis (CF) was localized to human chromosome 7q in 1985, and close markers, which are in linkage disequilibrium with the mutation, were later obtained (Estivill et al. 1987a,b). From the allelic association data, it appears that the great majority of CF chromosomes carry an identical mutation and are descended from an original founder; there is a second mutation at the same locus, which is present in carriers in Southern Italy and Spain. A considerable effort is now being made by several groups of investigators to determine the precise mutation causing CF. The biochemical defect is not known, and there is no carrier test.

As for most autosomal recessive conditions, families in which CF occurs usually have no previous history of the disease; when an affected child is born, there is a considerable physical, economic, and emotional burden to which each family reacts in an individual way (Kaback 1984). Many request prenatal diagnosis for subsequent pregnancies, usually because they wish a termination of pregnancy if the fetus is affected and occasionally to prepare better for family and medical care. There are also many requests for carrier testing for other family members, in particular siblings both of those affected and of the parents of CF children.

With a closely linked probe, it is possible to offer these diagnoses to families provided that a DNA sample from the affected individual is available to establish phase. The accuracy of the prediction of status, whether for carriers or for those affected, depends on the closeness of the marker to the mutation. For the close markers (CS.7 and KM-19), we estimate the frequency of recombination as 1 in 1000 meioses. Using these markers, as well as other markers that had been isolated previously and that are ~1 cM (1% recombination) from the CF locus, many prenatal diagnoses and carrier tests have been carried out (Farrall et al. 1986a,b, 1987; Beaudet et al. 1988).

The two closest probes, KM-19 and CS.7, have been adapted for use with the polymerase chain reaction (PCR) technique by the Houston investigators and ourselves (Feldman et al. 1988; Lench et al. 1988; Williams et al. 1988b). The probes J3.11 and MET have also been adapted for PCR (K. Klinger; A. Beaudet; both pers. comm.). This has led to advantages in shortening the time taken for prenatal diagnosis, allowing the use of smaller samples and of less invasive methods of obtaining DNA, and in the potential for new approaches to carrier testing in the community.

PCR for Rapid Prenatal Diagnosis
Prenatal diagnosis using DNA is now often carried out on samples of fetal chorion obtained in the first trimester (Williamson et al. 1981); this permits diagnosis before quickening, which is more acceptable to most pregnant women than mid-trimester diagnosis, and permits a termination of pregnancy (if requested) to be carried out by a clinically safer procedure. However, there is often a 2-week wait for prenatal diagnosis by Southern blot analysis. We have been able to use PCR with sequences of the closely linked probe CS.7 to offer rapid prenatal diagnosis to families in our hospital; the sample is taken in the morning, the analysis carried out, and the result is available to the clinician to give to the family on the same day (Williams et al. 1988b). In these circumstances, the family can make choices quickly if they wish, or they can have time to think if they prefer.

We have also used PCR with CS.7 to make use of a Guthrie spot (blood spot taken in the first week of life for diagnosis of phenylketonuria) for DNA analysis; the dried blood spot was 17-years old from a child who had died some years ago (Williams et al. 1988a). Fortunately, pathologists often save such samples; in our Regional center, blood spots have been collected for the past 20 years and are available for clinical testing if requested by a family.

PCR with Buccal Cells
Since PCR is so sensitive, we have attempted to use buccal cells from a mouthwash for analysis and have found these to be per-

fectly satisfactory (Lench et al. 1988). The person is asked to rinse the mouth with 10 ml of water, the rinse is collected in a sterile plastic centrifuge tube, centrifuged at 3000 rpm for 5 minutes, the pellet is taken up in 0.5 ml of water, boiled for 3 minutes, cooled, and centrifuged again. One tenth of the supernatant is then used directly as template in the PCR mix. This also does away with the need to collect blood samples, desirable to minimize concern about AIDS among those giving and taking samples.

Since PCR does not involve the use of radioactive isotopes, expensive matrices of nitrocellulose or nylon, sterile syringes and needles, or trained clinical staff, it is inexpensive. We find this of importance in the United Kingdom, where resources allocated to the Health Services are being increasingly examined for cost effectiveness. The cost of Southern blot analysis is low when considering a family at high risk (when there has been a previously affected child), but expense becomes a major concern when considering the implications of community-wide carrier testing.

It is possible to amplify more than one DNA sequence using PCR; C. Coutelle (G.D.R. Academy of Sciences) recently simultaneously amplified sequences linked to CF and Duchenne's muscular dystrophy. He also successfully analyzed the genotype of a single human oocyte as a model for diagnosis, using single cells from a preimplantation embryo after in vitro fertilization (C. Coutelle, pers. comm.). This has also been demonstrated to be feasible for single human sperm (Li et al. 1988).

Community Carrier Testing
What of community-wide carrier testing? What is offered now, what will be possible in the future, and will it be wanted? Because the existing probes CS.7 and KM-19 are in considerable disequilibrium with CF, it is already possible to modify risk for those with no previous history of CF in their families (Farrall et al. 1987; Beaudet et al. 1989). Those who do not have the haplotype usually associated with CF on either chromosome have their risk modified from 1/20 to ~1/150. Those with a single CF haplotype rise to a 1/8 risk, and those with CF-associated alleles on both chromosomes have a risk of 1/4 of being carriers.

Offering alterations, however, in risk of this type is fraught with problems; most of those counseled are not skilled in statistical risk analysis, and a great deal of explanation is re-

71

quired. Perhaps more important, those at high risk still only have a maximum chance of 1/64 of having a child with CF, and there is little further that can be offered. By the very nature of the data, DNA analysis cannot be performed. Although the CF mutation is almost always associated with one haplotype (known as haplotype B), this still occurs more commonly on non-CF chromosomes than on those with the defect. Because of this, we only offer these risk alterations when someone who is the partner of a known carrier (e.g., a parent who remarries or a tested sibling) requests counseling.

We have attempted, however, to assess whether carrier testing would be welcomed by people without previous knowledge or family history of CF. To our surprise, a pilot survey carried out by medical students in London showed that over 80% of those asked in a structured interview stated that they wished to know their carrier status for CF. Although there will be no way of knowing whether this intention will be reflected in practice when the mutation has been defined, the availability of PCR, its ease of application using cells obtained noninvasively, and its relative cheapness indicate that testing will be feasible should it be requested by large numbers of people.

ACKNOWLEDGMENTS
This work was supported by generous grants from the Cystic Fibrosis Research Trust and the Medical Research Council. We have used references directly relevant to the application of PCR to CF; there are many groups that are applying similar techniques to other inherited diseases.

REFERENCES
Beaudet, A.L., G.L. Feldman, S.D. Fernback, G.J. Buffone, and W.E. O'Brien. 1989. Linkage disequilibrium, cystic fibrosis and genetic counseling. *Am. J. Hum. Genet.* **44**: 319.
Beaudet, A., J. Spence, M. Montes, W. O'Brien, X. Estivill, M. Farrall, and R. Williamson. 1988. Experience with new DNA markers for the diagnosis of cystic fibrosis. *N. Engl. J. Med.* **318**: 50.
Estivill, X., P. Scambler, B. Wainwright, K. Hawley, P. Frederick, M. Schwartz, M. Baiget, J. Keer, R. Williamson, and M. Farrall. 1987a. Patterns of polymorphism and linkage disequilibrium for cystic fibrosis. *Genomics* **1**: 257.
Estivill, X., M. Farrall, P. Scambler, G. Bell, K. Hawley, N. Lench, G. Bates, H. Kruyer, P. Frederick, P. Stanier, E. Watson, R. Williamson, and B. Wainwright. 1987b. A candidate for the cystic fibrosis locus isolated by selection for methylation-free islands. *Nature* **326**: 840.

Farrall, M., X. Estivill, and R. Williamson. 1987. Indirect cystic fibrosis detection. *Lancet* **II:** 156.

Farrall, M., P. Scambler, K.W. Klinger, K. Davies, C. Worrall, R. Williamson, and B. Wainwright. 1986a. Cystic fibrosis carrier detection using a linked gene probe. *J. Med. Genet.* **23:** 295.

Farrall, M., C. Rodeck, P. Stanier, W. Lissens, E. Watson, H.-Y. Law, R. Warren, M. Super, P. Scambler, B. Wainwright, and R. Williamson. 1986b. First trimester prenatal diagnosis of cystic fibrosis with linked DNA probes. *Lancet* **I:** 1402.

Feldman, G., R. Williamson, A. Beaudet, and W. O'Brien. 1988. Prenatal diagnosis of cystic fibrosis by DNA amplification for detection of KM-19 polymorphism. *Lancet* **II:** 102.

Kaback, M. 1984. Attitudes toward prenatal diagnosis of cystic fibrosis among parents of affected children. In *Cystic fibrosis: Horizons* (ed. D. Lawson), p. 15. Wiley, London.

Lench, N., P. Stanier, and R. Williamson. 1988. A simple non-invasive method to obtain DNA for gene analysis. *Lancet* **II:** 1356.

Li, H., U.B. Gyllensten, X. Xui, R.K. Saiki, H.A. Erlich, and N. Arnheim. 1988. Amplification and analysis of DNA sequences in single human sperm and diploid cells. *Nature* **335:** 414.

Williams, C., L. Weber, and R. Williamson. 1988a. Guthrie spots for DNA-based carrier testing in cystic fibrosis. *Lancet* **II:** 693.

Williams, C., C. Coutelle, R. Williamson, F. Loeffler, J. Smith, and A. Ivinson. 1988b. Same-day first-trimester antenatal diagnosis for cystic fibrosis by gene amplification. *Lancet* **II:** 102.

Williamson, R., J. Eskdale, D.V. Coleman, M. Niazi, F.E. Loeffler, and B.M. Modell. 1981. Direct gene analysis of chorionic villi: A possible technique for first trimester antenatal diagnosis of haemoglobinopathies. *Lancet* **II:** 1125.

Multiplex Amplification for Diagnosis of Duchenne's Muscular Dystrophy

J.S. Chamberlain,[1] R.A. Gibbs,[1] J.E. Ranier,[1] N.J. Farwell,[1] and C.T. Caskey[1,2]

[1]Institute for Molecular Genetics and [2]Howard Hughes Medical Institute, Baylor College of Medicine, Houston, Texas 77030

The polymerase chain reaction (PCR) is a powerful method for the analysis of specific nucleic acid sequences (Saiki et al. 1988). This technique is useful for amplifying small quantities of genomic DNA or cDNA, which can then be used for DNA cloning, sequence analysis, mutation detection, or polymorphism identification (Kogan et al. 1987; Veres et al. 1987; Wong et al. 1987; Lee et al. 1988). Prior uses of this procedure have involved the amplification and subsequent examination of sequences known previously to exist in a sample. We have explored the use of PCR procedures to distinguish the presence or absence of a particular nucleic acid sequence. Such an application should permit rapid and sensitive assays for exogenous DNA or RNA sequences introduced into a sample via infection, transfection, or contamination, as well as detect the loss of sequences resulting from chromosomal or vector-based deletions. We have also determined that PCR procedures can be modified to permit the simultaneous amplification of at least nine separate DNA sequences, facilitating the analysis of multiple sequences from a single specimen. This multiplex amplification technique has proven useful for the diagnosis of a majority of genetic lesions leading to Duchenne's muscular dystrophy (DMD).

As described elsewhere in this volume, PCR has been used extensively to detect mutations leading to other genetic disorders, and these simple and rapid techniques provide an attractive alternative to *DMD* mutation detection procedures currently in use. With the use of recently isolated cDNA clones for the *DMD* gene (Burghess et al. 1987; Koenig et al. 1987), Southern blot analysis can detect deletions in the DNA of up to 60% of DMD patients (den Dunnen et al. 1987; Koenig et al. 1987; Baumbach et al. 1989) and is useful for haplotyping and linkage analysis in the great majority of DMD families (Hejtmancik et al. 1986; Ward et al. 1989). However, the large size ($>2 \times 10^6$ bp) and number of exons (>60) of this gene render Southern blot analysis a time-consuming, tedious, and expensive method for diagnosis.

Successful application of PCR to DMD diagnosis requires that the method be capable of reliably detecting deletions. PCR primers flanking DMD exon 17 were utilized to demonstrate that deletions were detectable in the DNA of patients and could be used successfully for prenatal diagnosis (Chamberlain et al. 1989). However, the 35-kb average size of *DMD* gene introns would necessitate separately amplifying each exon for complete detection of the deletions observed to arise frequently in this disorder via new mutation. Although heterogeneously distributed, *DMD* gene deletions tend to be of such a large size that a subset of only nine exons was observed to be deleted from the DNA of approximately 50% of all patients analyzed (Koenig et al. 1987; Baumbach et al. 1988; Chamberlain et al. 1988). Therefore, genomic clones, each containing one of these exons, were isolated, and sequence data from flanking introns were used to synthesize PCR primers (Fig. 1). By sequentially modifying reaction conditions every time a new set of primers was incorporated into the assays, a method was developed that permitted simultaneous amplification of all nine regions in a single reaction (Fig. 2A).

We have previously described reaction conditions under which six of the DMD regions were coamplified (Chamberlain et al. 1988). In those assays, 70% of the *DMD* gene deletions were detectable using template DNA isolated from lymphoblasts of affected males. Similar results were obtained with DNA extracted from cultured amniotic fluid cells and chorionic villus specimens (CVS) dissected of decidual tissue, indicating that the method could be used for prenatal mutation detection. The use of CVS DNA raised a primary concern about applying sensitive amplification procedures in a negative assay, i.e., the successful detection of deletions required that amplification not produce visibly detectable signals. Contamination of template DNA with nondeleted DNA or amplification of sequences not specifically targeted could lead to spurious results, causing misinterpretation of the data and derivation of a misdiagnosis. We have demonstrated that under appropriate conditions amplification from nontargeted DNA does not occur and that levels of contaminating maternal DNA at up to 5% of the total

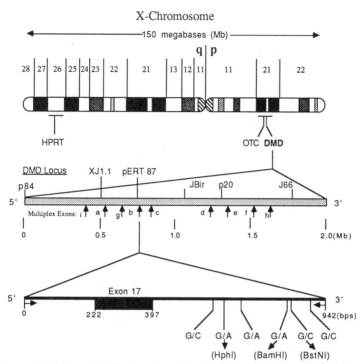

Figure 1 Schematic illustration of the X chromosome and *DMD* locus. (*Top*) Location of the *DMD* gene on the X chromosome. (*Middle*) *DMD* gene illustrating the relative location of nine exons amplified in Fig. 2A (lanes *a–i*), the approximate size of the locus, and the location of RFLP-detecting genomic clones. (*Bottom*) Expanded view of exon 17, illustrating the location of PCR primers (→) as well as six DSPs and the polymorphic base changes observed.

will not lead to false positive results (Chamberlain et al. 1988). Tests are under way to ensure that the nine-plex reaction (Fig. 2A) will be equally reliable.

As currently applied, the multiplex amplification assays are uninformative for the ~40% of DMD patients that do not display intragenic deletions. Modification of the techniques to permit polymorphism detection for haplotyping and linkage analysis would permit prenatal diagnosis or carrier detection for virtually all DMD families. Toward this goal, informative DNA sequence polymorphisms (DSPs) are being sought near each of the exons targeted in the DMD multiplex amplification reactions. Analysis of approximately 600 bp at the 5′ end of intron 17 has revealed the presence of six separate DSPs (Fig. 1).

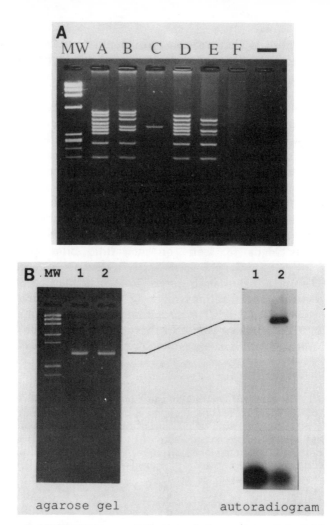

Figure 2 (*A*) Multiplex amplification of nine regions of the *DMD* gene for deletion detection. Primer sets flanking nine exons of the human *DMD* gene were pooled and used to amplify template DNA from six unrelated male DMD patients (lanes *A–F*). (Lane *A*) No deletion detected; (lanes *B–F*) partial or total deletion detected; (−) no template DNA added to the reaction; (lane *MW*) *Hae*III-digested φX174 DNA. The location of the nine exons are indicated in Figure 1. (*B*) COP analysis for allele determination at *DMD* gene intron 17. A 942-bp amplification product flanking exon 17 was analyzed via COP to determine the allele of the 5′ most G/C DSP of intron 17 (Fig. 1). Two reactions were performed, each containing both ASO primers specific for the DSP, only one or the other of which was labeled in each reaction. (*Left*) Ethidium-bromide-stained agarose gel showing COP products. (*Right*) Autoradiogram of the gel. Only one COP product was labeled, demonstrating that the G allele was present in the amplified DNA.

A variety of conventional methods are available to assay these six alleles for haplotyping. The DSPs that occur in restriction endonuclease recognition sequences are detectable as restriction-fragment-length polymorphisms (RFLPs) via digestion of amplified DNA with the appropriate restriction endonuclease. Alternatively, any DSP can be distinguished via hybridization of amplified DNA with allele-specific oligonucleotides (ASO) (Kogan et al. 1987).

DSPs can be directly detected during amplification by the use of competitive oligonucleotide priming (COP) (Gibbs et al., this volume). ASO primers specific for the G/C polymorphism nearest to exon 17 (Fig. 1) have been used successfully to determine alleles during COP amplification of this region of the gene (Fig. 2B). The procedure requires only a few hours of additional time for COP amplification beyond that required for standard PCR procedures, and the entire reaction can be performed on an automatic thermocycler. This aspect of COP reactions eliminates the need for hybridization analysis of amplified DNA, and aside from reagent addition and gel electrophoresis, the procedure is automated. Furthermore, the use of fluorescent primers will permit analysis of amplification products using an automated DNA-sequencing apparatus. COP analysis has proven successful not only with a single amplified region, but also when the entire nine-plex amplification reaction was used. These results suggest that automated multiplex amplification reactions can be used to detect deletions and determine haplotypes simultaneously, enabling informative linkage data to be derived from reactions that do not reveal deletions.

Integration of COP reactions into the multiplex amplification procedures for haplotyping throughout the *DMD* gene will require that informative DSPs be identified near a number of the exons targeted by the reactions. Observation of six DSPs within 600 bp of exon 17 suggests that polymorphisms may be quite common in introns (Fig. 1). Informative DSPs should occur much more frequently than RFLPs, since there is no requirement for overlap with restriction endonuclease recognition sequences, and yet known RFLPs can also be readily detected by COP analysis.

We have demonstrated that the use of multiplex amplification can permit rapid analysis of deletions located heterogeneously throughout the 2-Mb *DMD* gene. The availability of additional sequence and polymorphism data should permit modification of the methods for simultaneous deletion and haplotype detection. The procedures are not limited to the *DMD* locus but should be generally applicable to a number of molecular systems. For example, multiplex amplification can allow simultaneous analysis of entire genes or sets of genes for sequence analysis or mutation detection, would permit rapid screening of blood samples for numerous exogenous nucleic acid samples, and could be used for the amplification of repetitive sequence elements. Combined with methods for polymorphism detection, these procedures should also be generally applicable to linkage analysis, gene mapping, and forensic applications of DNA fingerprinting.

ACKNOWLEDGMENTS
We thank Elsa Perez and Lillian Tanagho for preparation of the manuscript. J.S.C. is supported by a grant from the Texas Higher Education Coordinating Board Advanced Technology Program. R.A.G. is a recipient of the Robert G. Sampson Distinguished Research Fellowship from the Muscular Dystrophy Association. This work was supported by a Task Force on Genetics grant from the Muscular Dystrophy Association and the Texas Higher Education Coordinating Board Advanced Technology Program.

REFERENCES

Baumbach, L.L., J.S. Chamberlain, P.A. Ward, N.J. Farwell, and C.T. Caskey. 1989. Molecular and clinical correlations of deletions leading to Duchenne muscular dystrophy. *Neurology* (in press).

Burghess, A.H.M., C. Logan, X. Hu, B. Belfal, R.G. Worton, and P.N. Ray. 1987. A cDNA clone for the Duchenne/Becker muscular dystrophy gene. *Nature* 328: 434.

Chamberlain, J.S., R.A. Gibbs, J.E. Ranier, P.N. Nguyen, and C.T. Caskey. 1988. Deletion screening of the Duchenne muscular dystrophy locus via multiplex DNA amplification. *Nucleic Acids Res.* 16: 11141.

Chamberlain, J.S., J.E. Ranier, J.A. Pearlman, N.J. Farwell, R.A. Gibbs, P.N. Nguyen, D.M. Muzny, and C.T. Caskey. 1989. Analysis of Duchenne muscular dystrophy mutations in mice and humans. *UCLA Symp. Mol. Cell. Biol. New Ser.* 93: 951.

den Dunnen, J.T., E. Bakker, E.G. Klein Breteler, P.L. Pearson, and G.J.B. van Ommen. 1987. Direct detection of more than 50% of the Duchenne muscular dystrophy mutations by field inversion gels. *Nature* 329: 640.

Hejtmancik, J.F., S.G. Harris, C.C. Tsao, P.A. Ward, and C.T. Caskey. 1986. Carrier diagnosis of Duchenne muscular dystrophy using restriction fragment length polymorphisms. *Neurology* 36: 1553.

Koenig, M., E.P. Hoffman, C.J. Bertelson, A.P. Monaco, C.C. Feener, and L.M. Kunkel. 1987. Complete cloning of the Duchenne muscular dystrophy (DMD) cDNA and preliminary genomic organization of the DMD gene in normal and affected individuals. *Cell* **50:** 509.

Kogan, S.C., M. Doherty, and J. Gitschier. 1987. An improved method for prenatal diagnosis of genetic diseases by analysis of amplified DNA sequences. *N. Engl. J. Med.* **317:** 985.

Lee, C.C., X. Wu, R.A. Gibbs, R.G. Cook, D.M. Muzny, and C.T. Caskey. 1988. Generation of cDNA probes directed by amino acid sequence: Cloning of urate oxidase. *Science* **239:** 1288.

Saiki, R.K., D.H. Gelfand, S. Stoffel, S.S. Scharf, R. Higuchi, G.R. Horn, K.B. Mullis, and H.A. Erlich. 1988. Primer directed enzymatic amplification of DNA with a thermostable DNA polymerase. *Science* **239:** 487.

Veres, G., R.A. Gibbs, S.E. Scherer, and C.T. Caskey. 1987. The molecular basis of the sparse fur mouse mutation. *Science* **237:** 415.

Ward, P.A., J.F. Hejtmancik, J.A. Witkowski, L.L. Baumbach, S. Gunnell, J. Speer, P. Hawley, S. Latt, U. Tantravahi, and C.T. Caskey. 1989. Prenatal diagnosis of Duchenne muscular dystrophy: Prospective linkage analysis and retrospective dystrophin cDNA analysis. *Am. J. Hum. Genet.* **44:** 270.

Wong, C., C.E. Dowling, R.K. Saiki, R.G. Higuchi, H.E. Erlich, and H.H. Kazazian. 1987. Characterization of β-thalassemia mutations using direct genomic sequencing of amplified single copy DNA. *Nature* **330:** 384.

Molecular Diagnosis of the Lesch-Nyhan Syndrome Using the Polymerase Chain Reaction

R.A. Gibbs,[1] P.-N. Nguyen,[2] L.J. McBride,[3] S.M. Koepf,[3] and C.T. Caskey[1,2]

[1]Baylor College of Medicine and [2]Howard Hughes Medical Institute
Houston, Texas 77030
[3]Applied Biosystems, Foster City, California 94404

The impact of techniques that are based on the polymerase chain reaction (PCR) (Saiki et al. 1985; Mullis and Faloona 1987) on the prospects for molecular diagnosis of human genetic disease has been tremendous. We have focused our efforts on the development of methods for the identification of mutations leading to the Lesch-Nyhan (LN) syndrome, an X-linked, genetically lethal disease that results from severe deficiency of the purine salvage enzyme hypoxanthine-guanine phosphoribosyltransferase (HPRT) (Lesch and Nyhan 1964; Kelley and Wyngaarden 1983; Stout and Caskey 1988). Prior to the invention of PCR, the identification of altered DNA sequences in a single LN patient was a formidable task that could not have been considered for routine diagnostic applications. With PCR amplification and the adaptation of automated DNA sequencing techniques, it is now possible to define rapidly the precise nucleotide alterations causing the disease in the majority of cases (Gibbs et al. 1989b). The DNA sequence alterations that are identified in the affected males can then be used as the basis for simplified tests for carrier status or early diagnosis of other family members. The advent of PCR has therefore made possible a strategy of LN diagnosis via direct detection of the mutant alleles, overcoming previous diagnostic problems associated with the molecular heterogeneity of LN mutations (Yang et al. 1984) and the difficulty in performing DNA linkage studies at the human *HPRT* locus (Stout and Caskey 1988).

The human *HPRT* gene is relatively large (≥44 kb) when compared to the *HPRT* peptide coding region (651 bases) (Patel et al. 1986), and the majority of LN patients produce sufficient

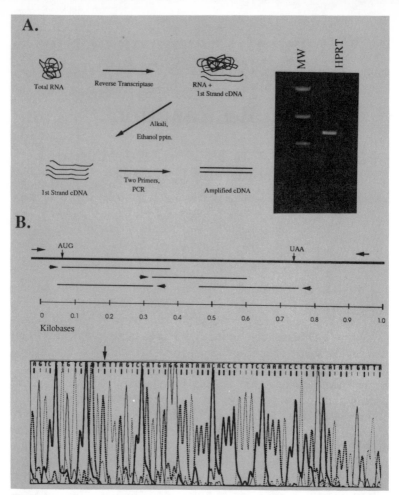

Figure 1 Identification of new mutations in human *HPRT* cDNA by PCR amplification and direct automated fluorescent DNA sequencing. (*A*) Total cellular RNA is copied by random hexamer-primed reverse transcription, and *HPRT* cDNA is amplified by PCR using specific oligonucleotide primers. A 938-bp product that contains the entire HPRT peptide coding sequence appears homogeneous when analyzed by agarose gel electrophoresis and ethidium bromide staining (lane *mw* standards are *Hae*III-digested φX174 DNA). (*B*) Sequence analysis of amplified *HPRT* cDNA. The solid line in the diagram represents the *HPRT* cDNA. PCR primers used in *A* are shown by arrows above the solid line. The position of binding of four fluorescent DNA sequencing primers and the length of sequence read from each are shown below the solid line. The chromatogram beneath shows a computer output following analysis of amplified *HPRT* cDNA from a LN patient with a single DNA base deleted (normal sequence CATAATTA, patient sequence CATATTA). The thick broken line indicates the "A" signal, the solid thick line indicates the "C" signal, the thin broken line indicates the "T" signal, and the solid thin line indicates the "G" signal.

HPRT mRNA to allow the synthesis of amplified *HPRT* cDNA via PCR. Thus, DNA sequence analysis of the *HPRT* cDNA represents the most convenient technique for searching for subtle gene alterations in LN patients. In the present method for *HPRT* mutation detection, total cellular RNA is isolated from transformed lymphoblast cultures. A cDNA copy is then generated from the RNA by randomly primed reverse transcription, and the entire 651-bp *HPRT*-coding region is PCR-amplified using a single set of HPRT-specific oligonucleotide primers. The PCR generates a DNA fragment that can be clearly resolved by ethidium-bromide-stained agarose gel electrophoresis (Fig. 1A).

The PCR-amplified *HPRT* cDNA is then directly subjected to DNA sequence analysis. To facilitate the sequencing, the double-stranded PCR product is first converted into a mixture containing an excess of one DNA strand. A small aliquot ($\leq 1\%$) of the PCR product is used as a template in a "single-strand" reaction (SSR) under conditions that are identical to a usual PCR except that only a single primer is used. The products of the SSR serve as superior DNA sequencing templates, because the sequencing primer does not have to compete with the opposing DNA strand when binding to the DNA template. This procedure was first described by Gyllensten and Erlich (1988), who performed single "asymmetric" PCRs with unequal mixtures of PCR primers to generate a single-stranded product in one reaction. The separation of the conventional double-stranded PCR and the SSR has the advantages of allowing the success of the initial reaction to be monitored and for the PCR products to serve as template for several different SSRs. The latter is a useful advantage as the permissible length of PCR amplification is currently much greater than the number of DNA bases that can usually be resolved in a single DNA sequencing reaction. In our hands, the separation of the PCR and the SSRs also generally increases the reliability of the overall analysis.

The DNA sequence analysis is performed using oligonucleotide primers that are labeled at their 5′ terminus and are complementary to bases that are between the PCR primers. The strategy generally overcomes the problems of spurious priming of nonspecifically amplified DNA that can arise when the same primers are used for PCR and DNA sequencing. The sequencing may be carried out by manual methods using ^{32}P-labeled DNA primers or on an automated DNA sequencing machine by using fluorescent oligonucleotides (Smith et al. 1986; Connell et al. 1987). An example of mutation detection by automated fluorescent DNA sequencing is shown in Figure 1B.

The automation of the DNA sequence analysis considerably simplifies the overall assay. In addition to the predictable advantages of no radioactivity and the automated entry of individual sequences, it is usually possible to obtain more sequence information per reaction than with manual methods. We currently employ three fluorescent DNA primer sets to enable the complete coverage of the *HPRT* peptide coding region (Fig. 1B). The potential of the automated direct sequencing has not yet been fully realized, and we predict that complete coverage of the *HPRT*-coding region will soon be possible with just one or two dye-labeled primer sets.

The major disadvantage of the fluorescent sequencing method is the necessity for preparation of suitable dye-labeled oligonucleotides. One alternative to the construction of new primers for each different fragment of DNA to be analyzed is described elsewhere (see McBride et al., this volume). In the case of the HPRT sequence analysis, there is sufficient interest for repeated determinations to justify the effort required to synthesize a batch of the dye-labeled primers. The methods for primer construction are described elsewhere (Gibbs et al. 1989b) and are relatively straightforward, provided the user is equipped with a flexible oligonucleotide synthesis machine.

We have now analyzed a total of 20 altered human *HPRT* cDNAs by direct DNA sequencing, revealing a wide range of mutation types (Table 1). The changes include single DNA base substitutions, DNA deletions, insertions, and errors in RNA splicing. Two of the single DNA base substitutions have been

Table 1 Classes of Mutations Identified in Human *HPRT* cDNA

Mutation	Number
Single DNA base substitutions	
amino acid substitution	7
premature chain termination	2
Single-base insertion	1
Deletion within a single exon	4
Deletion with insertion	1
RNA splicing mutations	5
Total	20

reported previously, and thus our studies confirm those findings (Davidson et al. 1988a,b). The remaining single-base substitutions encode amino acid changes that must inactivate the enzyme or premature translation termination codons that would result in a truncated HPRT protein. With a single exception, the cDNA base insertions and deletions disrupt the mRNA reading frame to prevent translation of downstream sequences. One 3-base deletion that does not cause a frameshift must affect either a functionally critical amino acid or a region that is important for the normal structure of the protein.

In the majority of cases, the identification of altered cDNA sequence directly infers the alteration in the genomic DNA of the affected individual. The identification of LN carriers and the diagnosis of future cases in those families is possible following the construction of synthetic oligonucleotides that perfectly match either the normal or mutant alleles. The most common procedure that is currently used is to radiolabel the allele-specific oligonucleotides (ASOs) and carry out sequential hybridizations to DNA fragments containing the mutation (Kidd et al. 1983). Stringent washing of the hybrids distinguishes the binding of perfectly matched oligomers from hybrids that contain a mismatch because of the mutation, and the identification of the oligonucleotides that bind at the most stringent conditions infers the mutation sequence. ASO probing has provided a relatively simple method for LN family analysis (Gibbs et al. 1989b), but the procedure is still somewhat cumbersome as it requires both hybridization to filter supports and careful control of the washing conditions.

As an alternative to ASO probing, we have developed a procedure for detection of known single DNA base changes based on competition between two oligonucleotide primers for DNA synthesis (Gibbs et al. 1988a). Competitive oligonucleotide priming (COP) is based on the strong preference of a DNA template for a perfectly matched oligonucleotide primer, relative to a primer with a single DNA base difference. ASO primers are prepared with the DNA base differences generally at or near their center. The primers are then mixed together with an aliquot of PCR-amplified genomic DNA, and the mixture is heated and cooled so the perfectly matched oligonucleotide may bind and be extended from the 3′ terminus by a DNA polymerase. A third "common" primer that binds nearby to the opposite DNA strand allows PCR amplification of a fragment that contains the successfully competing COP primer.

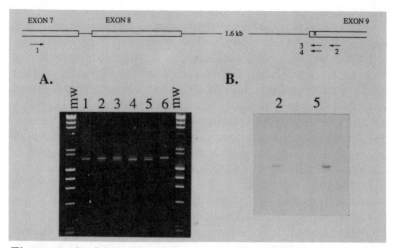

Figure 2 Analysis of a mutant human *HPRT* allele by COP. (*Top*) The position of hybridization of oligonucleotide primers for the analysis of a G→A transition that was first identified in amplified *HPRT* cDNA from a LN patient (cell line RJK 1727). The position of the mutation is indicated by X. The exons and introns are not to scale. Primers 1 and 2 were complementary to regions of exons 7 and 9, respectively, and were used to amplify an approximately 1.95-kb region containing the mutation from genomic DNA of RJK 1727 and a normal unrelated male. Oligonucleotide primers 3 and 4 were constructed to be complementary to the normal and mutant sequences, respectively. Aliquots of the reactions containing the amplified gene fragments were used to initiate COP reactions with primers 1, 3, and 4. In each case, duplicate COP reactions were performed with either primer 3 or primer 4 radiolabeled. The COP reactions were cycled through ten rounds of heating, annealing, and polymerization, and were analyzed by agarose gel electrophoresis (*A*) and autoradiography (*B*). The autoradiograph shows the incorporation of the radiolabeled primers when they are perfectly complementary to the DNA templates. (*A*) (Lane *mw*) Molecular weight standards; (lane *1*) PCR product generated by primers 1 and 2 and DNA from RJK 1727; (lane *2*) COP product, using amplified RJK 1727 DNA template with primer 3 (perfect match) radiolabeled; (lane *3*) COP product, using amplified RJK 1727 DNA template with primer 4 (mismatch) radiolabeled; (lane *4*) COP product, using normal male amplified DNA template, with primer 3 (mismatch) radiolabeled; (lane *5*) COP product, using normal male amplified DNA template, with primer 4 (perfect match) radiolabeled; (lane *6*) PCR product generated by primers 1 and 2 and DNA from a normal male.

When duplicate reactions are carried out with the allele-specific primer radiolabeled, a greater than 100-fold difference in signal intensity can result from the incorporation of a correctly matched DNA primer rather than a mismatched primer (Fig. 2).

The manipulations currently required for the COP procedure are thermocycling reactions, agarose gel electrophoresis, and autoradiography. COP is therefore more simple to perform than ASO analysis and better suited for the rapid genotyping of the LN families. The procedure is also well suited for automation. Modifications, such as fluorescent labeling of the COP primers and the addition of a biotin residue to the 5' terminus of the common primers, could further facilitate the method by allowing rescue of a PCR-amplified fragment that has incorporated the successfully competing, differentially labeled primer by an avidin-bound support. In this case, the final analysis of the COP products would not require gel electrophoresis and could be monitored by measurement of the radioactive or fluorescent incorporation of the support matrix. Such refinements may eventually lead to the complete automation of DNA base difference detection for genetic disease diagnosis and the analysis of other DNA sequence polymorphisms in complex genomes (Caskey 1987; Gibbs and Caskey 1989).

In five LN patients, the cDNA alterations involve exon boundaries and therefore are candidate RNA splicing mutations. As the cDNA sequence does not directly infer the gene mutations in these cases, the elucidation of the primary defects will require analysis of DNA that flanks the individual HPRT exons. This limits the ultimate utility of the *HPRT* cDNA sequencing method for LN diagnosis. In the near future, as the DNA sequence of the *HPRT* gene is better defined (Edwards et al. 1988), it will be possible to construct PCR primers to amplify each of the individual HPRT exons and the surrounding RNA splicing signals directly from genomic DNA samples. It is likely that the nine exon-specific amplification reactions may be carried out simultaneously (Chamberlain et al. 1988 and this volume) and that the mixture may be used to prime separate DNA sequencing reactions for each of the exons. In that case, a choice of cDNA or genomic sequencing will be available for the analysis of any human *HPRT* mutation.

ACKNOWLEDGMENTS

We thank John Belmont, Jeff Chamberlain, and Grant Mac-Gregor for valuable discussions; Andrew Civitello, Donna Muzny, and Sukeshi Vaishnev for technical assistance; John Moyer for preparation of the black and white plot of our DNA sequencing data; and the many physicians who have referred their Lesch-Nyhan patients to us. R.A.G. is a recipient of the

Muscular Dystrophy Association's Robert G. Sampson Distinguished Research Fellowship, and C.T.C. is an investigator of the Howard Hughes Medical Institute. This work was supported by Department of Human Services grant DK-31428 and Welch Foundation grant Q-533.

REFERENCES

Caskey, C.T. 1987. Disease diagnosis by recombinant DNA methods. *Science* **236**: 1223.

Chamberlain, J.S., R.A. Gibbs, J.A. Ranier, P.N. Nguyen, and C.T. Caskey. 1988. Deletion screening of the DMD locus via multiplex genomic DNA amplification. *Nucleic Acids Res.* **16**: 11141.

Connell, C., S. Fung, C. Heiner, J. Bridgham, E. Chakerian, E. Heron, B. Jones, S. Menchen, W. Mordan, M. Raff, M. Recknor, L. Smith, J. Springer, S. Woo, and M. Hunkapiller. 1987. Automated DNA sequence analysis. *Biotechniques* **5**: 342.

Davidson, B.L., T.D. Palella, and W.N. Kelley. 1988a. Genetic basis of hypoxanthine-guanine phosphoribosyltransferase deficiency in a patient with Lesch-Nyhan syndrome (HPRT$_{FLINT}$). *Gene* **68**: 85.

Davidson, B.L., M. Pashmforoush, W.N. Kelley, and T.D. Palella. 1988b. Human hypoxanthine-guanine phosphoribosyltransferase: A single nucleotide substitution in cDNA clones isolated from a patient with Lesch-Nyhan syndrome (HPRT$_{MIDLAND}$). *Gene* **63**: 331.

Edwards, A., J. Stegemann, C. Schwager, H. Voss, P. Rice, T. Kristenson, W. Ansorge, and C.T. Caskey. 1988. Automated DNA sequencing of the human HPRT gene: Use of the sequence for mutant characterization. *Am. J. Hum. Genet.* **43**: A182.

Gibbs, R.A. and C.T. Caskey. 1989. The application of genetic probes in epidemiology. *Ann. Rev. Pub. Health* **10**: 27.

Gibbs, R.A., P.N. Nguyen, and C.T. Caskey. 1989a. Identification of single DNA base differences by competitive oligonucleotide priming. *Nucleic Acids Res.* (in press).

Gibbs, R.A., P.N. Nguyen, L.J. McBride, S.M. Keopf, and C.T. Caskey. 1989b. Identification of mutations leading to the Lesch-Nyhan syndrome by automated direct DNA sequencing of *in vitro* amplified cDNA. *Proc. Natl. Acad. Sci.* **86**: (in press).

Gyllensten, U.B. and H. Erlich. 1988. Generation of single stranded DNA by the polymerase chain reaction and its application to direct sequencing of the HLA-DQA locus. *Proc. Natl. Acad. Sci.* **85**: 7652.

Kelley, W.N. and J.B. Wyngaarden. 1983. Clinical syndromes associated with hypoxanthine phosphoribosyltransferase deficiency. In *The metabolic basis of inherited disease*, 5th edition (ed. J.B. Stanbury et al.), p. 1115. McGraw Hill, New York.

Kidd, V.J., R.B. Wallace, I. Itakura, and S.L.C. Woo. 1983. Alpha-antitrypsin deficiency detection by direct analysis of the mutation in the gene. *Nature* **304**: 230.

Lesch, M. and W.L. Nyhan. 1964. A familial disorder of uric acid metabolism and central nervous system function. *Am. J. Med.* **36**: 561.

Mullis, K. and F.A. Faloona. 1987. Specific synthesis of DNA *in vitro* via a polymerase catalyzed chain reaction. *Methods Enzymol.* **155:** 35.

Patel, P.I., P.E. Framson, C.T. Caskey, and A.C. Chinault. 1986. Fine structure of the human hypoxanthine phosphoribosyltransferase gene. *Mol. Cell. Biol.* **6:** 393.

Saiki, R.K., F. Scharf, F. Faloona, K.B. Mullis, G. Horn, H.A. Erlich, and N. Arnheim. 1985. Enzymatic amplification of β-globin genomic sequences and restriction site analysis for diagnosis of sickle cell anemia. *Science* **230:** 1350.

Smith, L.M., J.Z. Sanders, R.J. Kaiser, P. Hughes, C. Dodd, C.R. Connell, C. Heiner, S.B.H. Kent, and L.E. Hood. 1986. Fluorescence detection in automated DNA sequence analysis. *Nature* **321:** 674.

Stout, J.T. and C.T. Caskey. 1988. The Lesch-Nyhan syndrome: Clinical, molecular and genetic aspects. *Trends Genet.* **4:** 175.

Yang, T.P., P.I. Patel, A.C. Chinault, J.T. Stout, L.G. Jackson, B.M. Hildebrand, and C.T. Caskey. 1984. Molecular evidence for new mutation at the HPRT locus in Lesch-Nyhan patients. *Nature* **310:** 412.

Human von Willebrand's Disease: Analysis of Platelet mRNA by Polymerase Chain Reaction

D. Ginsburg,[1,2] B.A. Konkle,[1,2,4] J.C. Gill,[3]
R.R. Montgomery,[1,2] P.L. Bockenstedt,[2]
T.A. Johnson,[1,2] and A.Y. Yang[1,2]

[1]Howard Hughes Medical Institute and Departments of Human
Genetics and [2]Internal Medicine, University of Michigan
Medical School, Ann Arbor, Michigan 48109-0650
[3]The Blood Center of Southeastern Wisconsin and the Medical College
of Wisconsin, Milwaukee, Wisconsin 53233

von Willebrand factor (vWF) is a multimeric plasma glyco-protein that plays an integral role in blood coagulation. vWF synthesis and posttranslational processing is complex, result-ing in the assembly of a series of high-molecular-mass multi-mers, ranging up to 20×10^6 daltons. vWF mRNA is 8.7 kb (en-coding 2813 amino acids), and the human vWF gene spans greater than 150 kb, interrupted by a minimum of 36 introns. vWF gene expression is exquisitely tissue specific and is thought to occur only in vascular endothelial cells and the bone marrow megakaryocyte (Ginsburg et al. 1985; Sadler et al. 1986; Ruggeri and Zimmerman 1987).

von Willebrand's Disease (vWD) is the most common in-herited bleeding disorder in man with estimated prevalence as high as 1% (Rodeghiero et al. 1987). At least 20 distinct clinical and laboratory subtypes and variants of the disease have been described. Although their molecular defects are not understood, they all share laboratory evidence for either a quantitative (type I) or qualitative (type II) abnormality of the vWF mole-cule (Ruggeri and Zimmerman 1987).

Aside from the clear autosomal dominant inheritance in the majority of cases of vWD, little is known about the genetics of this disorder. Although large gene deletions have been docu-mented in a few cases by Southern blot analysis (Shelton-Inloes et al. 1987; Ngo et al. 1988), the vWF gene appears grossly nor-

---

[4]Present address: Thomas Jefferson University, Philadelphia, Pennsyl-vania 19107.

93

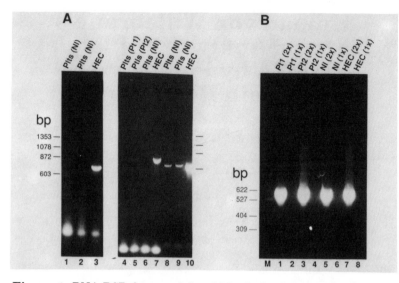

Figure 1 RNA PCR from peripheral blood platelets. (A) (Lanes 1–3) RNA PCR performed on approximately 1 μg of total RNA using oligo(dT) as reverse transcriptase (RTase) primer, and PCR primers designed to amplify a 709-bp fragment near the 3' end of vWF mRNA (7010–7718 [Bonthron et al. 1986a]); Templates for lanes 1 and 2 were total platelet RNA prepared from two different normals, and for lane 3 HUVEC RNA was used. (Lanes 4–10) Platelet RNA PCR, using a specific RTase primer, and PCR primers designed to amplify segments of 726 and 631 bp in a region near the middle portion of vWF mRNA (4622–5347 [Bonthon et al. 1986a]). RNA templates were platelet RNA from type IIA vWD patients 1 and 2 (lanes 4 and 5), platelet RNA from normal individuals (lanes 6,8, and 9), and HUVEC RNA (lanes 7 and 10). (B) Of the first round PCR products shown in lanes 4–7 in A above, 1 μl was used as template for a second round of PCR using the primers from lanes 8–9 (internal to the first round PCR primers). Of the reaction products from this second round PCR, 5 μl are shown in lanes 1, 3, 5, and 7, respectively. An equivalent dilution of singly amplified material not subject to the second round of PCR is shown in lanes 2, 4, 6, and 8, respectively.

mal in the majority of vWD patients (Ginsburg et al. 1985; Shelton-Inloes 1987; Ngo et al. 1988). Further molecular characterization of vWD has been difficult because of the large size and complexity of the vWF gene and the lack of a readily available source of patient vWF mRNA.

Amplification of vWF mRNA by Polymerase Chain Reaction
An RNA polymerase chain reaction (PCR) procedure was developed using small amounts of total cellular RNA as template for

first strand cDNA synthesis with this material, then used directly as template for PCR (Mullis and Faloona 1987; Saiki et al. 1988). By this approach, several segments of vWF mRNA were successfully amplified from peripheral blood platelet RNA (Fig. 1A). Optimal results were achieved using nested primers, that is, two consecutive rounds of PCR with the primer set for the second round located internal to the first set (Fig. 1B). An intense band was obtained from platelets indistinguishable from that obtained with endothelial cells, suggesting that under these conditions the amount of RNA template is not limiting, with the extent of amplification determined by consumption of reagents (or accumulation of inhibitory by-products) rather than by template quantity.

These results clearly demonstrate the presence of vWF mRNA in peripheral blood platelets, anucleate cell fragments derived from the marrow megakaryocyte, which are generally thought to contain little intact mRNA in normal individuals (Kieffer et al. 1987). The PCR product loaded in each lane in Figure 1B corresponds to the amount of platelet RNA obtained from approximately 1 µl of peripheral blood. In addition to the analysis of vWD, platelet RNA PCR should prove generally useful for the study of a variety of platelet-specific proteins. A similar approach has recently been used for the study of the platelet surface protein GPIIIA (Newman et al. 1988).

Detection of Missense Mutations in Type IIA vWD
Plasma of type IIA vWD patients has been observed to contain an increased amount of a 176-kD proteolytic fragment that has been localized toward the carboxy-terminal end of vWF (Berkowitz et al. 1987). The nested sets of PCR primers described above (Fig. 1B) were designed to amplify a segment of vWF mRNA in the region of the predicted amino terminus of this fragment. RNA PCR was performed on platelet RNA prepared from each of two patients with type IIA vWD (Fig. 1B). The PCR products were then subcloned into M13 using synthetic restriction sites incorporated into the PCR primers (Mullis and Faloona 1987) and sequenced, either as individual M13 clones or pools of >100 clones. By using pools of multiple clones, only authentic sequence differences will be detected, and random sequence errors due to the PCR procedure (Saiki et al. 1988) should not be apparent.

A different, single nucleotide change was observed in each patient. For patient 1, a GTC codon is changed to GAC, resulting in substitution of an aspartic acid for Val-844 of mature vWF. For patient 2, a CGG codon is converted to TGG, substituting a tryptophan for Arg-834. For patient 1, only the mutant sequence was detected, whereas in patient 2 both normal and mutant sequences were seen with approximately equal frequency. The normal vWF sequence contains two *Bst*EII restriction sites in this region located 35 bp apart. The first of these is destroyed by the mutation seen in patient 2 and the second by that observed in patient 1 (Fig. 2A). *Bst*EII digestion of PCR product from this region is shown in Figure 2B. Only the abnormal, large fragment (500 bp) is seen in patient 1, confirming the absence of the normal sequence from his mRNA. For patient 2, both the normal and mutant patterns are seen, also consistent with the direct sequence analysis.

Analysis of Genomic vWF Sequences by PCR

Analysis of genomic vWF sequences is complicated by the presence of a vWF pseudogene that will coamplify by PCR. A single-base difference between these genes in the target region creates an *Nco*I site in the pseudogene that is absent from the authentic vWF gene. To eliminate sequences arising from the pseudogene, genomic DNA templates were predigested with *Nco*I prior to amplification. Genomic DNA PCR analysis of 16 individuals from 9 type II vWD families (including 6 with type IIA) and 42 normal controls all gave only the normal *Bst*EII-digestion pattern. For the family of patient 2, the presence of the mutation was seen in the proband and in three family members affected with type IIA vWD and was absent from five other family members. All four affected individuals were seen to be heterozygous for the mutation, consistent with the RNA PCR data.

Identification of a vWF Allele Silent at the mRNA Level

As noted above, for patient 1 only the mutant sequence was observed by RNA PCR, a puzzling finding given the autosomal dominant nature of type IIA vWD and the presumed heterozygous state of most patients. The genomic DNA PCR results for patient 1 (confirmed by direct sequence analysis) indicated that he is indeed heterozygous for his mutation and that the mutation is absent from his clinically normal mother and brother. The patient's father, who was unavailable for study, clinically has type IIA vWD and was presumably the source of the patient's mutant allele. Taken together with the RNA PCR

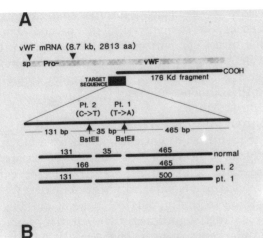

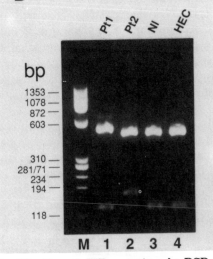

Figure 2 Identification of vWF mutations by PCR restriction digests. (A) The locations of the two type IIA mutations within the target PCR fragment are indicated. Each change eliminates a different *Bst*EII site (↑) present in the normal vWF sequence. The expected patterns for *Bst*EII digestion are shown. (B) Of the RNA PCR reactions from Fig. 1B, 5 μl was digested with *Bst*EII and fractionated on a 4% agarose gel. Position of size markers is shown in the left margin. Templates were patient 1 platelet RNA (lane *1*), patient 2 platelet RNA (lane *2*), normal platelet RNA (lane *3*), and HUVEC RNA (lane *4*).

results, these data indicate that the maternal vWF allele is either transcriptionally silent or gives rise to an improperly processed or unstable mRNA. Interestingly, patient 1 does have a clinically more severe bleeding disorder than his father, possibly explained by the coincidence of this second abnormal vWF allele. Such direct comparison of mRNA sequence to genomic DNA at a heterozygous nucleotide position may provide a general approach for determining the *cis* versus *trans* nature of a defect in mRNA production.

Expression of Mutant vWF

Each of the identified mutations was inserted into a cDNA expression vector containing full-length vWF sequences (Bonthron et al. 1986b). COS cells were transfected with normal, mutant, or control plasmid, and recombinant vWF was analyzed according to the methods of Wise et al. (1988). The expression construct containing the mutation from patient 1 resulted in marked disruption of the multimer pattern, closely resembling the characteristic pattern observed in type IIA patient plasma. Introduction of the patient 2 mutation produced a normal multimer pattern indistinguishable from that seen with wild-type vWF. These data suggest that the characteristic loss of high-molecular-weight multimers seen in type IIA vWD may occur through more than one mechanism.

The Molecular Basis of vWD is Heterogeneous

When mild forms are included, vWD appears to be one of the most common genetic disorders in man (Rodeghiero et al. 1987). vWD is clinically quite heterogeneous with over 20 subtypes defined, all with subtle phenotypic differences (Ruggeri and Zimmerman 1987). The data presented here demonstrate that even within these subtypes, additional molecular heterogeneity exists. In the two type IIA patients analyzed, a different mutation was identified in each, and neither mutation accounts for any of six additional type IIA cases screened by PCR. A defective vWF allele was identified that is silent at the mRNA level and would be expected to produce a quantitative abnormality in vWF. The frequency of this defect as the molecular basis for type I (the most common variant of vWD) must await further study. Using the approaches outlined here, it may be possible to define precisely the mutations accounting for a significant portion of vWD.

REFERENCES

Berkowitz, S.D., J. Dent, J. Roberts, Y. Fujimura, E.F. Plow, K. Titani, Z.M. Ruggeri, and T.S. Zimmerman. 1987. Epitope mapping of the von Willebrand factor subunit distinguishes fragments present in normal and type IIA von Willebrand disease from those generated by plasmin. *J. Clin. Invest.* **79:** 524.

Bonthron, D., E.C. Orr, L.M. Mitsock, D. Ginsburg, R.I. Handin, and S.H. Orkin. 1986a. Nucleotide sequence of pre-pro-von Willebrand factor cDNA. *Nucleic Acids Res.* **14:** 7125.

Bonthron, D.T., R.I. Handin, R.J. Kaufman, L.C. Wasley, E.C. Orr, L.M. Mitsock, B. Ewenstein, J. Loscalzo, D. Ginsburg, and S.H. Orkin. 1986b. Structure of pre-pro-von Willebrand factor and its expression in heterologous cells. *Nature* **324:** 270.

Ginsburg, D., R.I. Handin, D.T. Bonthron, T.A. Donlon, G.A.P. Bruns, S.A. Latt, and S.H. Orkin. 1985. Human von Willebrand factor (VWF): Isolation of cDNA clones and chromosomal localization. *Science* **228:** 1401.

Kieffer, N., J. Guichard, J.-P. Farcet, W. Vainchenker, and J. Breton-Gorius. 1987. Biosynthesis of major platelet proteins in human blood platelets. *Eur. J. Biochem.* **164:** 189.

Mullis, K.B. and F.A. Faloona. 1987. Specific synthesis of DNA in vitro via a polymerase-catalyzed chain reaction. *Methods Enzymol.* **155:** 335.

Newman, P.J., J. Gorski, G.C. White, S. Gidwitz, C.J. Cretney, and R.H. Aster. 1988. Enzymatic amplification of platelet-specific messenger RNA using the polymerase chain reaction. *J. Clin. Invest.* **82:** 739.

Ngo, K.Y., V.T. Glotz, J.A. Koziol, D.C. Lynch, J. Gitschier, P. Ranieri, N. Ciavarella, Z.M. Ruggeri, and T.S. Zimmerman. 1988. Homozygous and heterozygous deletions of the von Willebrand factor gene in patients and carriers of severe von Willebrand disease. *Proc. Natl. Acad. Sci.* **85:** 2753.

Rodeghiero, F., G. Castaman, and E. Dini. 1987. Epidemiological investigation of the prevalence of von Willebrand's disease. *Blood* **69:** 454.

Ruggeri, Z.M. and T.S. Zimmerman. 1987. von Willebrand factor and von Willebrand disease. *Blood* **70:** 895.

Sadler, J.E., B.B. Shelton-Inloes, J.M. Sorace, and K. Titani. 1986. Cloning of cDNA and genomic DNA for human von Willebrand factor. *Cold Spring Harbor Symp. Quant. Biol.* **51:** 515.

Saiki, R.K., D.H. Gelfand, S. Stoffel, S.J. Scharf, R. Higuchi, G.T. Horn, K.B. Mullis, and H.A. Erlich. 1988. Primer-directed enzymatic amplification of DNA with a thermostable DNA polymerase. *Science* **239:** 487.

Shelton-Inloes, B.B., F.F. Chehab, P.M. Mannucci, A.B. Federici, and J.E. Sadler. 1987. Gene deletions correlate with the development of alloantibodies in von Willebrand disease. *J. Clin. Invest.* **79:** 1459.

Wise, R.J., D.D. Pittman, R.I. Handin, R.J. Kaufman, and S.H. Orkin. 1988. The propeptide of von Willebrand factor independently mediates the assembly of von Willebrand multimers. *Cell* **52:** 229.

Problems with Fidelity of *Taq* DNA Polymerase in Searching for Mutations in the Human *PrP* Gene

D. Goldgaber

Department of Psychiatry, State University of New York at
Stony Brook, New York 11494

The prion protein (*PrP*) gene encodes a protein found in association with infectious agents causing Creutzfeldt-Jakob disease
(CJD) and kuru in humans and scrapie in animals. The nature
of these agents remains elusive. CJD is a rare neurodegenera-
tive disorder affecting approximately one person per million.
There is a very small number of cases of familial CJD. The genetic defect that determines familial CJD is unknown. Two
lines of evidence point to the *PrP* gene as a candidate gene for
familial CJD. First, the modified product of that gene is always
found in association with the infectious agent. Second, in
scrapie the changes in the coding region of the *PrP* gene were
found in various strains of mice having different scrapie incubation periods (Westaway et al. 1987). Recently, a rare *Msp*I
polymorphism was described in the coding region of the *PrP*
gene in a familial CJD case (Owen et al. 1988). To determine
whether a mutation in the *PrP* gene might be associated with
the familial CJD, the *PrP* gene of several CJD patients was sequenced. The human *PrP* gene encodes a 253-amino-acid
protein. The coding region of 759 bp is located on one exon. Figure 1 shows the outline of the experiments described in this
report. Genomic DNA was isolated from autopsied brain tissue
of CJD patients and coded. Two oliognucleotides corresponding
to the opposite ends of the *PrP* gene open reading frame (ORF)
were synthesized and used in a standard PCR reaction to
amplify the *PrP* ORF region from each coded DNA sample. The
oligonucleotides also contained unique restriction sites, which
were used for cloning of the amplified ORFs into pGEM-5z
plasmid (Promega Biotech).

The recombinant plasmids were transfected into *Escherichia coli*. Five to six individual colonies containing plasmids with

1. Genomic DNA

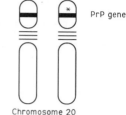

PrP gene

Chromosome 20

2. PCR

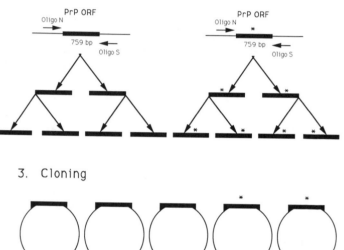

3. Cloning

4. Sequencing

Figure 1 Outline of the experimental procedure. Genomic DNA was isolated from postmortem brain tissues of familial and sporadic CJD patients and controls. The asterisks denote mutations. Thirty cycles of PCR reaction with *Taq* DNA polymerase were performed according to Cetus protocol. Two primers were utilized: oligo N, corresponding to the 5' end of the *PrP* ORF, and oligo S, corresponding to the 3' end of the *PrP* ORF. Each of these primers contained an artificial unique restriction site. The PCR amplified product was subcloned into pGEM-5z plasmid (Promega Biotech). The recombinant plasmid DNA was transfected into *E. coli* cells. Five to six individual colonies of cells transfected with recombinant DNA that was originated from one patient were separately grown in liquid cultures. Recombinant plasmid DNA was purified from each culture, and the *PrP* ORF region was then sequenced according to the Sanger method using specific primers.

PrP ORF of a single CJD patient were grown separately in liquid cultures. Each recombinant plasmid DNA was purified and sequenced by the Sanger method using specific primers. Thus, five to six sequences of the *PrP* ORF were obtained for each patient. Six sequences obtained in such a manner represented both alleles of the *PrP* gene with the probability of approximately 80%. If one of the alleles carried a mutation, then approximately half of the clones should have this mutation and the other half of the clones should not. Of the clones, 46 were sequenced and the samples were decoded. No mutation corresponding to the described *Msp*I polymorphism was found in our patients. A different unique mutation was found in two relatives with a familial CJD, and none of the other patients had this mutation. In one case, two out of five sequences, and in the other case, three out of five sequences contained this unique mutation. The distribution of sequences with the mutation corresponded to the expected distribution. However, the analysis of the *PrP* ORF sequences of some other patients revealed an unexpected distribution of a number of sequences with "mutations." For example, in one case one mutation was found in one of five sequences derived from the DNA of one patient. A different mutation was found in another sequence from the same patient. The other three sequences contained no mutations. This led to the conclusion that these mutations were, in fact, errors of the *Taq* DNA polymerase. Analysis of each step in the outlined procedure shows that the mutation detected in the final sequence may arise at each step involving synthesis of DNA: PCR, replication of the recombinant plasmids in *E. coli*, and sequencing. When the synthesis of DNA is an in vivo process, as in replication of recombinant plasmids in *E. coli*, the bacterial repair mechanism lowers the error rate to a level that is undetectable in the outlined experiments. The errors of the enzyme used for sequencing were identified by sequencing both strands of the recombinant plasmids. The choice, therefore, must be made between a true mutation that existed in the original genomic DNA and an error made by the *Taq* DNA polymerase. If the *Taq* DNA polymerase made an error during early cycles, the molecule with this error will be amplified, will be present in a large number of the amplified DNA molecules, and could be detected at the final sequencing step but could not be distinguished from the true mutation. To estimate maximum error rate of the *Taq* DNA polymerase in the setting of the outlined experiments with 30 cycles of PCR, the number of all questionable mutations (four) detected in all 46 sequenced *PrP* ORFs was divided by the total number of sequenced nucleotides (34,915). The error rate of approximately 1 in 9000 nucleotides was obtained. Surprisingly, this figure was practically the same as the error rate of *Taq* DNA polymerase estimated in the experiments involving only one cycle of PCR (Tindall and Kunkel 1988).

The conclusion here is that employing the PCR with *Taq* DNA polymerase in search for a mutation in a given gene is a valid approach. However, once a mutation in a gene is identified, it is critical to verify that the original DNA contains this mutation using an alternative method, for example, the allele-specific oligo method (Saiki et al. 1986). In addition, finding the same mutation in genomes of other members of the family will also confirm the existence of the identified mutation.

ACKNOWLEDGMENTS

I gratefully acknowledge the help of Drs. David M. Asher and Paul W. Brown in obtaining samples of CJD patients, Dr. Lev G. Goldfarb for the isolation of genomic DNA, and my students James W. Teener and Scott Lin for expert assistance in the project. The experiments were performed in the Carleton D. Gajdusek's Laboratory of Central Nervous System Studies, National Institute of Neurological and Communicative Disorders and Stroke, National Institutes of Health, Bethesda, Maryland.

REFERENCES

Owen, F., M. Poulter, R. Lofthouse, T.J. Crow, D. Risby, H.F. Baker, and R.M. Ridley. 1988. A rare Msp1 polymorphism in the human prion gene in a family with a history of early onset dementia. *Neurosci. Lett.* **32:** 553.

Saiki, R.K., T.L. Bugawan, G.T. Horn, K.B. Mullis, and H.A. Erlich. 1986. Analysis of enzymatically amplified β-globin and HLA-DQα DNA with allele-specific oligonucleotide probes. *Nature* **324:** 163.

Tindall, K.R. and T.A. Kunkel. 1988. Fidelity of DNA synthesis by the *Thermus aquaticus* DNA polymerase. *Biochemistry* **27:** 6008.

Westaway, D., P.A. Goodman, C.A. Mirenda, M.P. McKinley, G.A. Carlson, and S.B. Prusiner. 1987. Distinct prion proteins in short and long scrapie incubation period mice. *Cell* **51:** 651.

Specificity of the Polymerase Chain Reaction in the Study of α_1-Antitrypsin Deficiency

A. Graham,[1] C.R. Newton,[1]
S.J. Powell,[1] L.E. Heptinstall,[1] C. Summers,[1]
L. Brown,[1] R. Anwar,[1] K. Murray,[1]
A. Gammack,[1] R. Kennedy,[2] N. Kalsheker,[3]
and A.F. Markham[1]

[1]ICI Diagnostics, Gadbrook Park, Rudheath
Northwich, Cheshire CW9 7RA, United Kingdom
[2]Stobhill General Hospital, Glasgow G21 3UW, United Kingdom
[3]University of Wales College of Medicine, Department of
Biochemistry, Newport Road, Cardiff CF2 1SZ, United Kingdom

We have used the polymerase chain reaction (PCR) to examine mutations in the α_1-antitrypsin gene. Subjective observations on the reproducibility of the PCR of exons 3 and 5 after DNA preparation by different methods are discussed. Characterization of spurious amplification products occasionally observed in simultaneous amplification of these two exons has been performed. Amplification of this locus from blood spots on 21yo Guthrie PKU Test Cards is possible.

The fidelity of *Taq* polymerase has been examined by the sequencing of α_1-antitrypsin exon 3 and exon 5 PCR products after cloning into M13. Only a very low rate of mutation has been observed, consistent with the observations of other investigators. Direct sequencing of the PCR reaction products obtained by the amplification of α_1-antitrypsin exons 2, 3, 4, and 5 has been performed on a large number of individuals. In this situation, differences from the published gene sequences have not been observed to date in normal individuals. Reproducible diagnosis of heterozygotes and homozygotes for the common PI Z mutation has been performed. Tentative molecular characterization of three novel α_1-antitrypsin-deficient variants associated with lung and/or liver disease are described.

We have attempted to adopt the PCR with direct product sequencing (Kogan et al. 1987; Wong et al. 1987; Saiki et al. 1988) to the analysis of the PI Z and PI S alleles of α_1-anti-

trypsin (or α_1-protease inhibitor, PI). These single base mutations (342 Glu GAG to Lys AAG and 264 Glu GAA to Val GTA)
result in deficiency of α_1-antitrypsin and predispose
homozygotes to the development of childhood cirrhosis or pul-
monary emphysema. Two pairs of 30-mer PCR primers for exon
3 and exon 5 of the gene were prepared, which yield a 360-bp
PCR product (containing the potential S mutation) and a 220-
bp PCR product (containing the potential Z mutation), respectively (Newton et al. 1988).

PCR products of the expected size were obtained both when
individual exons were amplified separately and when both exons were amplified simultaneously. Occasionally, with certain
PCR cycle conditions an idiosyncratic band has been observed
at 124 bp. Our initial concern was that this might reflect
amplification within the recently described antitrypsin-related
gene (Bao et al. 1988). However, cloning and sequencing of this
band in M13 proved it to be a product formed by amplification
with the 5' exon 3 and 5' exon 5 primers. The central 64 bp
not attributable to either primer could not be identified in any
of the current DNA sequence data bases.

Amplification of DNA from Guthrie PKU Cards

There has been some speculation that an inhibitor of *Taq*
polymerase may copurify with human genomic DNA, particularly from blood (de Franchis et al. 1988; H. Erlich, pers.
comm.). We have used the rapid DNA isolation procedure of
Gitschier and co-workers (Kogan et al. 1987) without en-
countering these problems, provided that the blood samples (in
EDTA tubes) are subjected to a freeze/thaw cycle prior to the
workup of simply boiling in water (Fig. 1A). More recently, we
have attempted to amplify the exon 3 360-bp fragment in DNA
extracted from Guthrie cards of a variety of ages in a similar
manner (Lyonnet et al. 1988). Unexpectedly, the inclusion of a
small amount of SDS in the boiled samples significantly increases the yield of PCR products in our hands. The predicted

Figure 1 (A) Effect of the freeze/thaw cycle on the PCR products obtained by amplification of a 360-bp α_1-antitrypsin fragment from DNA
extracted by a simple boiling of pelleted blood cells in water. (B) Effect
of subsequent PCR of a 360-bp α_1-antitrypsin fragment of addition of
0.1% SDS to solution for simple boiling extraction of DNA from Guthrie card blood spots. (C) Analysis of a *Bst*EII polymorphism in a 360-
bp PCR fragment from α_1-antitrypsin exon 3, showing that purification of the PCR product prior to digestion is unnecessary.

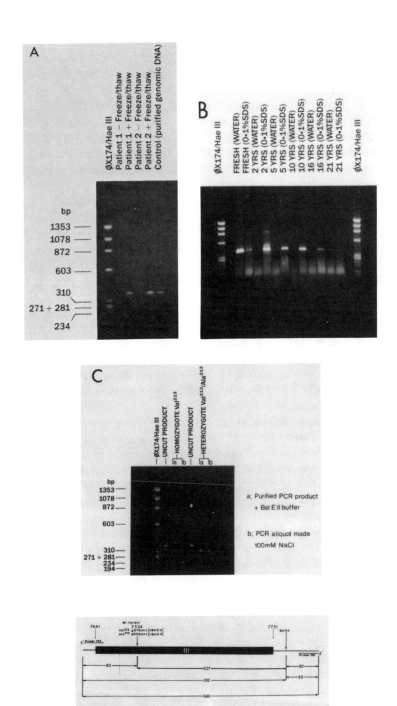

Figure 1 (*See facing page for legend.*)

amplification product has been obtained from Guthrie cards up to 21 years old (Fig. 1B). Direct analysis of polymorphic restriction sites within PCR fragments (Chehab et al. 1987; Feldman et al. 1988; Lench et al. 1988) has been applied for characterization of the *Bst*EIII polymorphism (Fig. 1C) arising as a result of the benign mutation 213 Val gGTGacc to Ala gGCGacc found in normal PI M1 alleles (Nukiwa et al. 1987b). PI Z alleles always contain also the 213 Ala form (Nukiwa et al. 1986).

Fidelity of Amplification with *Taq* Polymerase

Fidelity of the *Taq* polymerase (Dunning et al. 1988; Saiki et al. 1988; Tindall and Kunkel 1988) has been analyzed by cloning the 360-bp or 220-bp PCR products into M13 and sequencing individual clones. Overall, in 15 clones (4700 bp total sequenced) 11 were identical with published sequence, 1 contained an A→G plus an A→T mutation, 1 contained a G→A plus a G deletion, and 1 contained A→G, T→C, and A→G mutations. Thus seven errors were detected in 4700 bp total, in good agreement with estimates elsewhere (17 errors in 6692 bp, Saiki et al. 1988; 22 errors in 8000 bp, Dunning et al. 1988). An alternative approach to the assessment of polymerase accuracy (Tindall and Kunkel 1988) suggests that the error rate may be substantially lower.

Errors due to *Taq* polymerase have not been observed to date when direct sequencing is performed on bulk PCR products without cloning. This has been the case when examining exon 3 and exon 5 PCR products from numerous individuals in pedigrees with the MM, MZ, MS, SS, and ZZ genotypes. Reliable diagnosis of the various possible homozygotes and heterozygotes is achieved using either "nested" internal sequencing primers or the original PCR primers themselves in dideoxy sequencing reactions. The analysis has been extended to include exons 2 and 4, when the polymorphism leading to the PI, M1, and M2 isotypes is revealed (Nukiwa et al. 1988).

The ability to sequence coding (and intron) sequences rapidly of the α_1-antitrypsin gene provides a powerful tool for the characterization of patients and carriers in pedigrees presenting with suspected α_1-antitrypsin deficiency where the common PI S or PI Z variants are not responsible. A number of such "null" alleles have been described previously (Nukiwa et al. 1987a; Satoh et al. 1988). Using the PCR direct sequencing technique, we have characterized disease-associated mutations in three further PI variants: PI Null$_{Cardiff}$, PI M$_{Malton}$, and PI I.

108

REFERENCES

Bao, J., L. Reed-Fourquet, R.N. Sifers, V.J. Kidd, and S.L.C. Woo. 1988. Molecular structure and sequence homology of a gene related to α-1-antitrypsin in the human genome. *Genomics* **2**: 165.

Chehab, F.F., M. Doherty, S. Cai, Y.W. Kan, S. Cooper, and E.M. Rubin. 1987. Detection of sickle cell anemia and thalassaemias. *Nature* **329**: 293.

de Franchis, R., N.C.P. Cross, N.S. Foulkes, and T.M. Cox. 1988. A potent inhibitor of Taq polymerase copurifies with human genomic DNA. *Nucleic Acids Res.* **16**: 10355.

Dunning, A.M., P. Talmud, and S.E. Humphries. 1988. Errors in the polymerase chain reaction. *Nucleic Acids Res.* **16**: 10393.

Feldman, G.L., R. Williamson, A.L. Beaudet, and W.E. O'Brien. 1988. Prenatal diagnosis of cystic fibrosis by DNA amplification for detection of KM-19 polymorphism. *Lancet* **II**: 102.

Kogan, S.C., M. Doherty, and G. Gitschier. 1987. An improved method for prenatal diagnosis of genetic diseases by analysis of amplified DNA sequences: Application to haemophilia A. *N. Engl. J. Med.* **317**: 985.

Lench, N., P. Stanier, and R. Williamson. 1988. A simple non-invasive method to obtain DNA for gene analysis. *Lancet* **I**: 1356.

Lyonnet, S., C. Cailland, F. Rey, M. Berthelon, J. Frezal, J. Rey, and A. Munnich. 1988. Guthrie cards for detection of point mutations in phenylketonuria. *Lancet* **II**: 507.

Newton, C.R., N. Kalsheker, A. Graham, S.J. Powell, A. Gammack, J. Riley, and A.F. Markham. 1988. Diagnosis of α-1-antitrypsin deficiency by enzymatic amplification of human genomic DNA and direct sequencing of polymerase chain reaction products. *Nucleic Acids Res.* **16**: 8233.

Nukiwa, T., M.L. Brantly, F. Ogushi, G.A. Fells, and R.G. Crystal. 1988. Characterisation of the gene and protein of the common α-1-antitrypsin normal M2 allele. *Am. J. Hum. Genet.* **43**: 322.

Nukiwa, T., H. Takahashi, M. Brantly, M. Courtney, and R.G. Crystal. 1987a. α-1-Antitrypsin Null Granite Falls, a nonexpressing α-1 antitrypsin gene associated with a frameshift to stop mutation in a coding exon. *J. Biol. Chem.* **262**: 11999.

Nukiwa, T., K. Satoh, M.L. Brantly, F. Ogushi, G.A. Fells, M. Courtney, and R.G. Crystal. 1986. Identification of a second mutation in the protein-coding sequence of the Z type alpha 1-antitrypsin gene. *J. Biol. Chem.* **261**: 15989.

Nukiwa, T., M. Brantly, F. Ogushi, G. Fells, K. Satoh, L. Stier, M. Courtney, and R.G. Crystal. 1987b. Characterisation of the M1 (Ala^{213}) Type of α1 antitrypsin, a newly recognized, common "normal" α-1-antitrypsin haplotype. *Biochemistry* **26**: 5259.

Saiki, R.K., D.H. Gelfand, S. Stoffel, S.J. Scharf, R. Higuchi, G.T. Horn, K.B. Mullis, and H.A. Erlich. 1988. Primer-directed enzymatic amplification of DNA with a thermostable DNA polymerase. *Science* **239**: 487.

Satoh, K., T. Nukiwa, M. Brantly, R.I. Garver, M. Hofker, M. Courtney, and R.G. Crystal. 1988. Emphysema associated with complete absence of α-1-antitrypsin and of a stop codon in an α-1-antitrypsin coding exon. *Am. J. Hum. Genet.* **42**: 77.

Tindall, K.R. and T.A. Kunkel. 1988. Fidelity of DNA synthesis by the Thermus aquaticus DNA polymerase. *Biochemistry* **27:** 6008.

Wong, C., C.E. Dowling, R.K. Saiki, R.G. Higuchi, H.A. Erlich, and H.H. Kazazian. 1987. Characterization of β-thalassemia mutations using direct genomic sequencing of amplified single copy DNA. *Nature* **330:** 384.

Application of the Polymerase Chain Reaction to the Molecular Analysis of Mutations in Chinese Hamster Ovary (AS52) Cells

K.R. Tindall

Laboratory of Molecular Genetics, National Institute of
Environmental Health Sciences, Research Triangle Park
North Carolina 27709

Point mutations have been implicated in the etiology of a number of human genetic disorders as well as in the activation of some oncogenes. Little is known, however, about the mechanisms by which such mutations occur in mammalian cells. Determination of mechanistic pathways by generating mutational spectra using DNA sequence analysis has proven extremely successful in bacterial systems. Similar studies in mammalian cells, however, have been limited because of technical problems associated with the rapid isolation and analysis of target gene sequences (Moore et al. 1987).

We are analyzing both spontaneous and induced mutations in a transgenic Chinese hamster ovary (CHO) cell line, AS52, that carries a single copy of the bacterial guanine phosphoribosyltransferase (*gpt*) gene transfected and functionally integrated into the CHO genome (Tindall et al. 1984). The *gpt* gene is analogous to the mammalian hypoxanthine-guanine phosphoribosyltransferase (*hprt*) gene, and mutations at either locus can be isolated by selecting for resistance to the purine analog, 6-thioguanine (6TG[r]). We have overcome the problem of recovering mutant gene sequences from the mammalian genome by directly amplifying mutant *gpt* gene sequences using the polymerase chain reaction (PCR) technique (Saiki et al. 1985). The small size of the *gpt* structural gene (456 bp) provides for the convenient amplification of mutant *gpt* sequences from the CHO genome for subsequent DNA sequence analysis. The use of PCR allows us to generate sufficient quantities of DNA for sequence analysis without cloning, thus allowing the

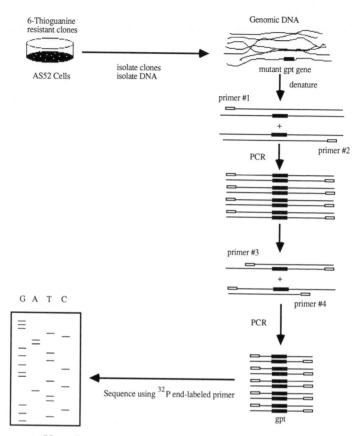

Figure 1 Use of nested-PCR amplification and DNA sequence analysis to generate mutant sequence spectra in AS52 cells. Genomic DNA from independent 6TG^r mutant clones is isolated, and 0.5–1 μg is used to amplify mutant *gpt* sequences by PCR. PCR amplification is initiated in the presence of oligonucleotide primers 1 and 2, dNTPs, and the *Taq* polymerase (Saiki et al. 1988). Following 15–30 rounds of synthesis, primers 1 and 2 are removed (see text), and primers 3 and 4 are added in excess. Primers 3 and 4 hybridize to sites on the amplified DNA that are internal with respect to primers 1 and 2. An additional aliquot of the *Taq* polymerase is added, and PCR is allowed to continue an additional 15–30 rounds. Before sequencing, primers 3 and 4 are removed from the reaction by centrifugation through Centricon 30 tubes. An aliquot of the PCR-generated product, corresponding to ~250 ng of target DNA is then sequenced by the dideoxy-sequencing technique using ^{32}P end-labeled primers and Sequenase (Tindall and Stankowski 1989). The sequencing reactions are subjected to electrophoresis on a 5.5% denaturing acrylamide gel. Overnight exposure to XAR-5 film yields autoradiograms that allow identification of the mutation. Protocols used for nested PCR amplification and sequencing of double-stranded DNA using end-labeled primers and Sequenase are available upon request.

rapid generation of point mutational spectra derived in mammalian cells. In addition, those mutants that do not yield a PCR-generated fragment of the appropriate size can be tentatively characterized as deletion mutants awaiting more detailed analyses by Southern blotting (see Tindall and Stankowski 1989).

PCR Amplification Using Nested Primers

Figure 1 illustrates the general approach used to analyze the AS52-6TG^r clones. Mostly, we have amplified mutant *gpt* sequences using a nested set of oligonucleotide primers. That is, by performing PCR for 15 to 30 rounds with a set of primers that flank the sequence of interest followed by an additional 15 to 30 rounds using a second set of flanking primers that are internal to the initial set, one can readily generate the sequence of interest with high yield and minimal amplification of secondary sites. Although optimal conditions for PCR can be determined using a single set of primers, we have found that the annealing and extension conditions vary with the primers and target sequence. The use of nested primers yields highly reproducible results with minimal variation between primers under our standard PCR conditions.

A limitation to the approach of nested amplification is that the initial set of primers must be removed before continuing PCR with the internal set. We have found two methods to be effective in eliminating the first set of primers. Usually, we have purified the PCR product by two rounds of centrifugation through Centricon 30 (Amicon) tubes. This process provides a sufficiently pure PCR product for subsequent amplification with the internal set of primers. Alternatively, we eliminate the Centricon 30 purification step by using limiting amounts of the initial primers for 20–30 rounds of PCR, followed by the addition of the internal set of primers and continued PCR amplification. This latter approach is similar in design to the generation of single-stranded DNA using limiting concentrations of one of the primers during PCR (Gyllensten and Erlich 1988) and yields high-quality PCR product with minimal manipulations.

Direct Sequence Analysis of the PCR Product

The heat-stable DNA polymerase from *Thermus aquaticus (Taq)* has rapidly become the most commonly used polymerase for PCR (Saiki et al. 1988). The *Taq* polymerase, however, is somewhat error-prone, yielding mutations in the synthesized DNA at a frequency of approximately $1/10^4$ nucleotides synthesized (Tindall and Kunkel 1988; see also Kunkel and Eckert, this volume). Such errors have important implications regarding the subsequent analysis of the PCR product, since any molecule cloned and sequenced following PCR may contain mutations that occurred during the amplification process.

If DNA sequence analysis is the ultimate goal, as is the case in generating mutant sequence spectra, then the contribution of errors occurring during amplification can be minimized by directly sequencing the PCR product. In our studies, we amplify a sample of mutant AS52 (0.5–1 µg) genomic DNA. In 1 µg of AS52 DNA, there are approximately 10^6 copies of the mutant *gpt* gene. Assuming the amplification process is not biased, then errors occurring even in the first round of PCR will be a small fraction of the final product. Thus, direct sequence analysis of the PCR product generates a genomic consensus sequence, and errors that occurred during PCR are simply not observed on the resulting autoradiogram. Two results support this statement. First, directly sequenced PCR-generated DNA from nonmutant AS52 cells results in the same sequence as reported previously for the functional *gpt* gene (Richardson et al. 1983). Second, in several cases, the sequence of the same 6TG[r] mutant AS52 clone, amplified by PCR in separate reactions, results in the identification of the same mutational change in both samples. Other mutations that might have arisen during PCR have not been observed in these samples.

In our studies, we sequence the double-stranded PCR generated product using ^{32}P end-labeled primers that hybridize to the *gpt* gene internal to the primers used in PCR (Fig. 1). We have optimized the conditions for sequencing the double-stranded PCR product with Sequenase (U.S. Biochemical Corp.), and different primers are used to sequence different regions of the *gpt* structural gene. Thus far, all 456 bases of the *gpt* structural gene have been sequenced for each mutant analyzed. Using Sequenase and the ^{32}P end-labeled primers, we consistently generate high-quality sequences of the *gpt* mutants, making the generation of single-stranded template or the cloning of the PCR product unnecessary.

Analysis of Spontaneous Mutations in AS52 Cells

A critical consideration in the development of any mechanistic arguments regarding induced mutations based on DNA sequence spectra is an understanding of the nature of the spontaneous mutations that arise within an assay system. Since any single mutant isolated for analysis may be spontaneous in origin, one should isolate induced mutants for analysis at mutant frequencies well above the observed spontaneous mutant frequency to be certain that most of the mutants analyzed are induced rather than spontaneous. In addition, analysis of the spontaneous mutations observed in a system will allow future comparisons with induced spectra (Adams and Skopek 1987). Mutational hotspots or mutational biases usually exist and can provide significant insight regarding mutational mechanisms.

We have recently defined the spontaneous spectrum for the *gpt* locus in AS52 cells (Tindall and Stankowski 1989) using both Southern blot analysis and PCR. 6TG^r mutants were isolated using a modified fluctuation analysis to ensure independence of each of the isolated mutant clones. In short, the results of two independent experiments indicate that most spontaneous AS52 mutations are deletions. In both experiments, approximately 50–60% of these deletions are detectable by Southern blot analysis. The remaining mutants (34/62) from experiment 2 either showed no detectable change on a Southern blot or generated a DNA fragment of appropriate size for *gpt* following PCR. All 34 mutants were PCR-amplified and sequenced. The results are presented in Table 1. Among the 34 mutants, most (21) were small deletions of either 2-, 3-, or 7-bases. Of these, a specific 3-base deletion is present in this spectrum representing a significant mutational hotspot. This 3-base deletion comprises 31% (19/62) of the spontaneous AS52 mutants analyzed. Notably, this same 3-base deletion is also significantly represented in the spontaneous spectrum generated at a chromosomally integrated *gpt* locus in mouse A9I2 cells (Ashman and Davidson 1987). The frequency among the spontaneous mutants in A9I2 cells, however, is somewhat lower, present in 21% of the spontaneous clones analyzed. Interestingly, the same 3-base deletion is not observed in *Escherichia coli* (Crosby et al. 1988), implying a role for chromatin structure or the mammalian replication complex in the generation of this specific deletion. We are currently in the process of evaluating possible pathways of generating the 3-base deletion at the *gpt* locus in AS52 cells at such high frequencies.

Of the remaining PCR-amplified mutants analyzed, one was

Table 1 DNA Sequence Spectrum of Spontaneous 6TG^r-AS52 Mutants

Mutations observed	Amino acid[a] change	Site[b]	No. of mutants observed at that site
Deletions			
7-base deletion	FS	368–374	1
3-base deletion	Δ Asp	366–371[c]	19
2-base deletion	FS	386–387	1
Insertions			
2-base insertion[d]	FS	261–262	1
Basepair substitutions[e]			
transitions			
GC→AT	Gly→Asp	242	1
AT→GC	Asp→Gly	275	1
transversions			
GC→TA	Glu→TAA (stop)	7	1
GC→TA	Trp→Leu	26	1
GC→CG	Gly→Arg	241	1
TA→AT	Val→Asp	377	1
AT→TA	Asp→Val	419	1
AT→TA	TAA (stop)→Leu	458	1
No change detected			4
Total			34

Adapted, with permission, from Tindall and Stankowski (1989).

[a]The amino acid change in the xanthine-guanine phosphoribosyltransferase protein as the result of the mutational change is presented. The 7- and 2-base deletions and the 2-base insertion result in a shift in the translational reading frame (FS). The 3-base deletion results in the loss of a single aspartic acid residue (Δ Asp).

[b]The mutational site is represented according to the numbering of the *gpt* gene (Richardson et al. 1983).

[c]There are four possible deletions of three adjacent bases in the region from base 366 to 371 that result in the same primary DNA sequence in the mutant.

[d]The insertion of two bases, 5'–GT–3', in the sense strand occurs between the T residue at base 261 and the G residue at base 262.

[e]Base substitutions are listed with the first base representing the base at the indicated site in the sense strand of the *gpt* structural gene. For example, there is a G in the sense strand at base 242 that is mutated to an A and reported above as a GC→AT transition at base 242.

a 2-base insertion, eight were base-pair substitutions, mostly resulting in nonconservative amino acid changes, and four did not contain mutations in the *gpt* structural gene (Table 1). These data provide a basis for comparing future induced mutational spectra derived in the AS52 cell line and suggest that spontaneous deletions may be a predominant mutational event in the CHO cell genome.

Future Directions

We intend to study further the pathways leading to spontaneous and induced deletions at the *gpt* locus in AS52 cells. Deletions and rearrangements may have serious genomic consequences, and there are few systems available to study such events in detail. The AS52 cell line seems to detect deletions that are not detected at other loci (e.g., *hprt*), perhaps providing a more accurate measure of the frequency of deletions occurring in the mammalian genome (Tindall and Stankowski 1987). We hope to exploit our ability to recover such deletions in the AS52 cell line to gain insight into the mechanistic pathways by which such deletions arise in mammalian cells.

In addition, we will continue to apply the PCR technique to the analysis of point mutations induced by a variety of agents. Most molecular studies in bacterial systems have focused on mutagens for which substantial data regarding DNA adducts are available. Following this lead, there are a number of agents of interest to this laboratory such as mitomycin C, psoralen plus near-UV light, and H_2O_2, which should provide insights about the roles of mammalian DNA repair, DNA adduction, and replication fidelity in the generation of mutations in mammalian cells. The ability to use PCR to amplify mammalian mutant sequences specifically has enabled us to generate a data base rapidly, allowing mechanistic hypotheses to be put forward and tested. The next few years should provide significant insights regarding mutational mechanisms in mammalian cells.

ACKNOWLEDGMENTS

I thank Dr. Leon Stankowski, Jr. (Pharmakon Research International, Inc., Waverly, Pennsylvania) for his help in the generation of the spontaneous AS52 mutants. In addition, the critical comments of Drs. Robert Tuveson and Jim Clark regarding this manuscript are appreciated. These studies were supported by the National Institute of Environmental Health Sciences and Pharmakon Research International, Inc.

REFERENCES

Adams, W.T. and T.R. Skopek. 1987. Statistical test for the comparison of samples from mutational spectra. *J. Mol. Biol.* **194:** 391.

Ashman, C.R. and R.L. Davidson. 1987. Sequence analysis of spontaneous mutations in a shuttle vector gene integrated into the mammalian chromosomal DNA. *Proc. Natl. Acad. Sci.* **84**: 3354.

Crosby, R.M., K.K. Richardson, T.R. Craft, K.B. Benforado, H.L. Liber, and T.R. Skopek. 1988. Molecular analysis of formaldehyde-induced mutations in human lymphoblasts and *E. coli. Environ. Mol. Mutagen.* **12**: 155.

Gyllensten, U.B. and H.A. Erlich. 1988. Generation of single stranded DNA by the polymerase chain reaction and its application to direct sequencing of the HLA-DQA locus. *Proc. Natl. Acad. Sci.* **85**: 7652.

Moore, M.M., D.M. DeMarini, F.J. de Serres, and K.R. Tindall, eds. 1987. *Mammalian cell mutagenesis.* Cold Spring Harbor Laboratory, Cold Spring Harbor, New York.

Richardson, K.K., J. Fostel, and T.R. Skopek. 1983. Nucleotide sequence of the xanthine guanine phosphoribosyl transferase gene of *E. coli. Nucleic Acids Res.* **11**: 8809.

Saiki, R.K., S. Scharf, F. Faloona, K.B. Mullis, G.T. Horn, H.A. Erlich, and N. Arnheim. 1985. Enzymatic amplification of β-globin genomic sequences and restriction site analysis for diagnosis of sickle cell anemia. *Science* **230**: 1350.

Saiki, R.K., D. Gelfand, S. Stoffel, S.J. Scharf, R. Higuchi, G.T. Horn, K.B. Mullis, and H.A. Erlich. 1988. Primer directed enzymatic amplification of DNA with a thermostable DNA polymerase. *Science* **239**: 487.

Tindall, K.R. and T.A. Kunkel. 1988. Fidelity of DNA synthesis by the *Thermus aquaticus* DNA polymerase. *Biochemistry* **27**: 6008.

Tindall, K.R. and L.F. Stankowski, Jr. 1987. Deletion mutations are associated with the differential-induced mutant frequency response of the AS52 and CHO-K1-BH4 cell lines. *Ban. Rep.* **28**: 283.

———. 1989. Molecular analysis of spontaneous mutations at the *gpt* locus in Chinese hamster ovary (AS52) cells. *Mutat. Res.* **220**: 241.

Tindall, K.R., L.F. Stankowski, Jr., R. Machanoff, and A.W. Hsie. 1984. Detection of deletion mutations in pSV2*gpt*-transformed CHO cells. *Mol. Cell. Biol.* **4**: 1411.

Sequence Determination of Point Mutations at the *HPRT* Locus in Mammalian Cells following In Vitro Amplification of *HPRT* cDNA Prepared from Total Cellular RNA

A.A. van Zeeland, H. Vrieling, J.W.I.M. Simons, and P.H.M. Lohman

Department of Radiation Genetics and Chemical Mutagenesis
Sylvius Laboratories, State University of Leiden
2333 AL Leiden, The Netherlands

The gene coding for hypoxanthine-guanine phosphoribosyl-transferase (HPRT) is often used in mutagenicity assays with mammalian cells. The gene is X-linked, and therefore in any diploid cell line, mutants can be selected for (van Zeeland and Simons 1976). In addition, HPRT mutants can also be obtained directly from lymphocytes of mouse (Jones et al. 1985) and human (Albertini et al. 1982; Morley et al. 1983) origins. Since there is a growing knowledge of the nature of DNA damage introduced by many chemical and physical mutagens, there is an increasing need to correlate the presence of a particular type of DNA damage with the nature of the ultimate DNA sequence change that occurred as a consequence of the DNA damage introduced. To obtain a spectrum of mutations within a particular gene that might be characteristic for a chemical or physical mutagen, it is necessary to analyze large numbers of mutants at the DNA sequence level.

The molecular analysis of the *HPRT* gene on the DNA sequence level is very time-consuming because the *HPRT* gene is a relatively large gene (34 kb in mouse and 42 kb in man). However, most single base-pair changes that affect the HPRTase activity are expected to occur in the coding sequence of the *HPRT* gene, which is only 654 bp long. Therefore, a methodology was developed, based on the polymerase chain reaction, which allowed rapid sequencing of base pair changes in the coding region of the *HPRT* gene (Vrieling et al. 1988).

General Principle of the Methodology

Total cytoplasmic RNA was isolated from mutants of V79 Chinese hamster cells, GRSL mouse lymphoma cells and human T lymphocytes. Total RNA (10–20 µg) was used for the synthesis of cDNA with Moloney murine leukemia reverse-transcriptase, using an oligonucleotide primer that can anneal 3′ of the stop codon of the HPRT coding sequence. A second primer is then added that can anneal to the newly synthesized DNA strand 5′ of the start codon of the *HPRT* gene. The HPRT coding region is subsequently amplified in 32 cycles of denaturation, primer annealing, and polymerization using *Taq* polymerase (Cetus Corp.). This procedure yields up to 5 µg of amplified DNA, depending on the primers used. The amplified DNA can be sequenced directly or following subcloning in M13-based sequencing vectors.

Selection of Primers

The yield of the amplification is very dependent on the characteristics of the two primers used. It is important that the melting temperature (T_m) of the primers are not too much different. The T_m can be calculated using the following formula:

$$T_m \, (^oC) = 4 \, (G + C) + 2 \, (A + T) - 5$$

Most of the primers that were used were 20-mer. When the amplified product was analyzed on 2% neutral agarose gels, sometimes a smaller fragment was observed in addition to the amplified HPRT fragment. This smaller fragment is sensitive to S1 nuclease and is also recognized by an *HPRT* cDNA probe. The band presumably contains amplified HPRT sequences but in a single-stranded (ss) form because the extension reaction from one of the two primers is more efficient. When the T_m of one of the primers was adjusted by increasing its length from 18 to 24 bases, the ssDNA band was not observed anymore, since now both primers had about an equal T_m. In the case of mouse and hamster cells, a set of primers could be selected that amplified the HPRT coding region satisfactorily. However, when a similar set of primers was used on human material, no signal or a very weak signal was observed. The reason for this was the presence of a palindrome structure that is located immediately 5′ to the start codon of the human HPRT coding sequence. When a primer was used that contained two mismatches by which it disrupts the palindrome, a strong signal

Table 1 Primers for the Amplification of the HPRT Coding Sequence

Primer	Species	Position	Sequence[a]
vrl-1	mouse	3'	5'-GGACTCCTCGTATTTGCAGA-3'
zee-1	mouse	5'	5'-GGCTTCCTCCTCCAGACCGCT-3'
vrl-16	hamster	3'	5'-GCAGATTCAACTTGAAtTCTCATC-3'
vrl-10	hamster	3'	5'-TCAACTTGAAtTCTCATC-3'
vrl-8	hamster	5'	5'-CCGCCAGCCGAtCGATTCCG-3'
vrl-17	human	3'	5'-CCAAACTCAACTTGAAtTCTCATC-3'
vrl-18	human	5'	5'-taCGCCGGaCGGaTCCGTT-3'

[a]Mismatches are indicated with small letters.

121

was obtained after amplification. The explanation is probably that disruption of this palindromic structure allows better annealing of the 5′ primer. The sequences of the primers presently used for the three species of human, mouse, and hamster are given in Table 1.

Errors Made by *Taq* Polymerase

A major part of the sequenced mutants were analyzed after cloning of amplified *HPRT* cDNA in M13 cloning vectors. This means that mistakes introduced by *Taq* polymerase during amplification become visible in the cloned DNA sequence. Therefore, only base changes that could be confirmed in independent M13 clones were assumed to be authentic mutations. Other base changes are most likely due to errors introduced by *Taq* polymerase. The error frequency of this enzyme was estimated to be about 1 in 900 sequenced bases after 32 cycles of amplification. The type of changes introduced by *Taq* polymerase itself is almost only the AT to GC transition.

UV-mutation Spectra in Normal and DNA-repair-deficient UV-sensitive V79 Chinese Hamster Cells

The influence of DNA repair on the molecular nature of mutations induced by UV light (254 nm) was investigated in UV-induced HPRT mutants from UV-sensitive Chinese hamster cells (V-H1) and its parental line V79 (Vrieling et al. 1989). The nature of point mutations in HPRT exon sequences was determined for 19 HPRT mutants of V79 and for 17 HPRT mutants of V-H1 cells by sequence analysis of in-vitro-amplified *HPRT* cDNA. The mutation spectrum in V79 cells consisted of single and tandem double base-pair changes, whereas in V-H1 cells three frameshift mutations were also detected. Among the mutants analyzed, four mutants were found in which one exon was missing from the amplified *HPRT* cDNA. This phenomenon is probably caused by a mutation in the splice acceptor site of the missing exon. All base pair changes in V-H1 mutants were due to GC to AT transitions. In contrast, in V79, all possible classes of base pair changes except the GC to CG transversion were present. In this group, 70% of the mutations were transversions. Since all mutations except one did occur at dipyrimidine sites, the assumption was made that they were caused by UV-induced photoproducts at these sites. This assumption allows the determination of the DNA strand of the

HPRT gene in which the damage causing the observed mutation is located. In V79 cells, 11 out of 17 base pair changes were caused by photoproducts in the nontranscribed strand of the *HPRT* gene. However, in V-H1 cells, which are completely deficient in the removal of pyrimidine dimers from the *HPRT* gene and which show a seven-times enhanced UV-induced mutation frequency, 10 out of 11 base pair changes were caused by photoproducts in the transcribed strand of the *HPRT* gene. The hypothesis is made that this extreme strand specificity in V-H1 cells is due to differences in fidelity of DNA replication of the leading and the lagging strand. Furthermore, it is proposed that in normal V79 cells two processes determine the strand specificity of UV-induced mutations in the *HPRT* gene, namely (1) preferential repair of the transcribed strand of the *HPRT* gene and (2) a higher fidelity of DNA replication of the nontranscribed strand compared with the transcribed strand.

Future Applications

Investigation of mutation spectra will be useful for many purposes. It may point out which lesions are mutagenic and what kind of influence chromosomal position and local DNA structure have on mutation induction. Furthermore, it can provide more insight in cellular processes such as (preferential) DNA repair and DNA replication. Since mutation spectra can also be obtained in vivo by isolating HPRT-deficient T lymphocytes from human blood, epidemiological studies may be performed that will possibly allow correlation of these spectra to occupational and environmental exposure of individuals to mutagenic agents. The mutagenic properties of indirect-acting mutagens, which have to be metabolized to become mutagenic, can be monitored following isolation of HPRT mutant lymphocytes from mouse blood.

ACKNOWLEDGMENTS

This research was sponsored by the J.A. Cohen Institute for Radiation Pathology and Radiation Protection, The Queen Wilhelmina Fund of the Netherlands, Euratom, and Royal Dutch Shell.

REFERENCES

Albertini, R.J., K.S. Castle, and W.R. Borcherding. 1982. T cell cloning to detect the mutant 6-thioguanine resistant lymphocytes present in human peripheral blood. *Proc. Natl. Acad. Sci.* **79:** 6617.

Jones, I.M., K. Burkhart-Schultz, and A.V. Carano. 1985. A method to quantify spontaneous and in vivo induced thioguanine-resistant mouse lymphocytes. *Mutat. Res.* **147:** 97.

Morley, A.A., K.J. Trainor, R. Seshadri, and R.B. Rydall. 1983. Measurements of in vivo mutations in human lymphocytes. *Nature* **302:** 155.

van Zeeland, A.A. and J.W.I.M. Simons. 1976. Linear dose-response relationships after prolonged expression times in V79 Chinese hamster cells. *Mutat. Res.* **35:** 129.

Vrieling, H., J.W.I.M. Simons, and A.A. van Zeeland. 1988. Nucleotide sequence determination of point mutations at the mouse HPRT locus using in vitro amplification of HPRT mRNA sequences. *Mutat. Res.* **198:** 107.

Vrieling, H., M.L. van Rooijen, N.A. Groen, M.Z. Zdzienicka, J.W.I.M. Simons, P.H.M. Lohman, and A.A. van Zeeland. 1989. DNA strand specificity for UV-induced mutations in mammalian cells. *Mol. Cell. Biol.* **9:** 1277.

HLA Class II Gene Polymorphism: Detection, Evolution, and Relationship to Disease Susceptibility

H.A. Erlich, R.K. Saiki, and U. Gyllensten

Human Genetics Department, Cetus Corporation
Emeryville, California 94608

The capacity of the polymerase chain reaction (PCR) (Saiki et al. 1985, 1988b; Mullis and Faloona 1987) to amplify a specific segment of genomic DNA has made it an invaluable tool in the study of polymorphism and evolution, as well as in the analysis of genetic susceptibility to disease. In all of these areas, a particular gene must be examined in a variety of individuals, either within a species, in different closely related species, or in patient and in healthy control populations. We have used PCR, initially with the Klenow fragment and more recently with *Taq* polymerase, to determine the allelic sequence diversity of the polymorphic HLA class II genes (*HLA-DRβ, HLA-DQα, HLA-DQβ*, and *HLA-DPβ*). Since serologically defined alleles at some of these loci have been associated with specific autoimmune diseases (e.g., insulin-dependent diabetes), we have compared the distribution of allelic DNA sequences at these class II loci in patients and in controls using oligonucleotide probes. The evolution of the class II polymorphism observed in the human population was examined by comparing the sequences of PCR-amplified class II gene segments from different individuals from a variety of primate species. Here, we discuss the evolution of the class II polymorphism and its relation to disease susceptibility, and we present a new and powerful method for the rapid genotyping of amplified DNA samples.

Class II Sequence Polymorphism
The *HLA-D* or class II genes are organized into three distinct regions, *HLA-DR, HLA-DQ*, and *HLA-DP*, each of which encode an α- and a β-glycopeptide. The association of these α- and β-chains forms a heterodimeric transmembrane protein expressed on a number of cell types, including B lymphocytes, macrophages, and activated T lymphocytes. These highly

polymorphic proteins bind peptide fragments of foreign antigen, and it is the resulting complex that is recognized by the T-cell receptor, leading to activation of the T lymphocyte. The polymorphism of class II genes is localized to the amino-terminal outer domain encoded by the second exon. Using PCR primers to conserved regions, we have amplified and sequenced the second exon of these class II loci from many different individuals, revealing a remarkable degree of allelic diversity. Some loci have as many as 25 alleles. These sequences were determined either by M13 cloning, followed by chain-termination sequencing of the purified single-stranded phage DNA, or by direct sequencing using the asymmetric primer method (Gyllensten and Erlich 1988) to generate single strands from the PCR.

Detection

The detection of the HLA class II polymorphism is valuable in the areas of individual identification, tissue-typing for transplantation, and genetic susceptibility to specific autoimmune diseases. The serologically defined allelic variants have been revealed to be heterogeneous at the DNA level. For example, the DRw6, DQw1 haplotype can contain any one of three different $DR\beta I$, three different $DR\beta III$, six different $DQ\beta$, and three different $DQ\alpha$ DNA-sequence-defined alleles. A variety of analytic techniques have been used to detect genetic variation in PCR-amplified DNA, including restriction enzyme digestion, RNase-A cleavage, denaturing gradient gel electrophoresis, sequence analysis, as well as allele-specific oligonucleotide (ASO) primers and ASO hybridization probes. For the analysis of amplified loci with many allelic variants, we have found that the use of nonradioactively labeled ASO or sequence-specific oligonucleotide (SSO) probes is the most general and convenient approach. (In some cases, the hybridization of an oligonucleotide probe does not uniquely specify an allele because the specific sequence is present in more than one allelic variant. In the absence of allele-specific sequences, a given allele is identified as a pattern of SSO probe binding.) This dot-blot procedure is a powerful and rapid genetic testing method and has been used for the diagnosis of sickle-cell anemia and β-thalassemia (Saiki et al. 1988a), as well as for $HLA\text{-}DQ\alpha$ genotyping for the identification of forensic samples (Bugawan et al. 1988a). However, for a locus with n alleles, each amplified sample must be immobilized on n membranes, and each membrane must be hybridized to one of n labeled probes. Thus, the procedural complexity of this approach is a function of the number of oligonucleotide probes required for complete genetic analysis. To address this problem, we have recently developed a "reverse dot-blot" procedure in which the oligonucleotide probe is immobilized on a membrane and hybridized to a labeled PCR product (Fig. 1) (Saiki et al. 1989). In this method,

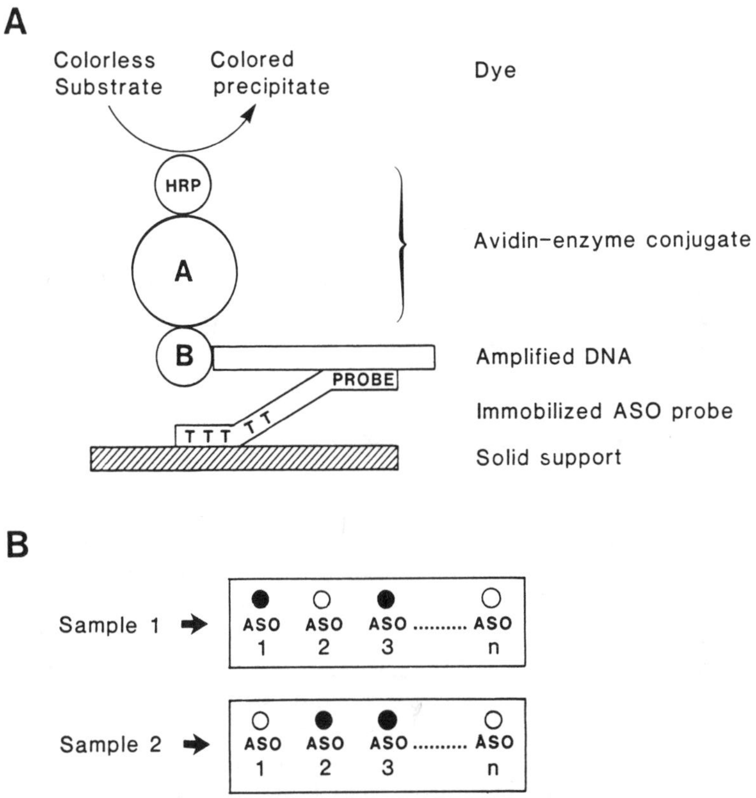

Figure 1 Schematic diagram of immobilized oligonucleotide probe detection of amplified DNA. (*A*) An allele-specific probe is "tailed" with a dT homopolymer and immobilized on a solid support. The amplified PCR product, which has incorporated a biotinylated primer, hybridizes to the probe. After washing away unbound DNA and primers, the biotinylated, amplified DNA binds an avidin horseradish peroxidase conjugate. The enzyme then converts a colorless dye into a colored precipitate. (*B*) A format for detecting specific alleles in samples of amplified DNA, e.g., different alleles for various HLA class II loci in samples from heterozygous individuals.

a panel of oligonucleotide probes is tailed with poly(dT), using terminal transferase, and UV-cross-linked to a nylon membrane. The PCR product, labeled during amplification by using biotinylated primers, is then hybridized to the immobilized array of oligonucleotide probes. The presence of the specifically bound PCR product is detected using a streptavidin horseradish peroxidase conjugate, which converts a colorless, soluble substrate to a colored precipitate. This method, described in detail by Saiki et al. (1989) has been applied to detection of a variety of β-thalassemia mutations and *HLA-DQα* alleles.

Evolution

The extensive polymorphism observed in the human class II loci could have been generated by recent (i.e., after speciation) mutation, recombination, or gene conversion, followed by selection for the newly arisen variants. Alternatively, an ancient (i.e., before speciation) polymorphism maintained by selection could have given rise to the observed allelic diversity. To test these hypotheses, we have amplified and sequenced the polymorphic second exon of the *DQα* locus from a number of individuals in a variety of primate species: the hominoids (e.g., humans, chimpanzees, and gorillas), some old world monkeys (e.g., baboons, rhesus, and langur), and some new world monkeys (e.g., capuchin and marmoset) (Gyllensten and Erlich 1989). Phylogenetic analysis of these sequences was carried out using the method of maximum parsimony. In general, a given human *DQα* allelic type (e.g., *DQA4*) is more closely related to its gorilla or chimpanzee allelic counterpart than it is to any other human allelic type (e.g., *DQA1*, *DQA2*, or *DQA3*). This pattern in which the sequences cluster by allelic types rather than by species suggests that most of the *DQα* allelic diversity in the contemporary human population was present in an ancestral species that gave rise to the human, chimpanzee, and gorilla lineages. Thus, at the *DQα* locus, the polymorphic alleles appear to be ancient (5–20 million years old, based on the known divergence times of the hominoids and old world monkeys) and have been maintained by selection (e.g., overdominance or frequency-dependent mechanisms). At the *DQβ* locus, the major allelic types (e.g., *DQB1*, *DQB2*, *DBB3*, and *DBB4*) also appear to be ancient, but here some subtypic diversification has occurred *after* speciation (U. Gyllensten et al., unpubl.).

Disease Susceptibility

The third area where PCR-based analysis of class II poly-morphisms has proved very valuable is the study of genetic susceptibility to HLA-associated diseases. A variety of diseases have been associated with specific serologically defined variants (e.g., insulin-dependent diabetes mellitus [IDDM] and DR3 and DR4). A disease-associated marker is simply one whose frequency is increased in patients relative to controls (e.g., 90% of IDDM patients are DR3 or DR4 compared with 40% of controls). PCR amplification facilitated the sequence analysis of class II alleles derived from patients and controls and the comparison of their distribution in the two populations by oligonucleotide probe hybridization (dot blot; see above). Since the serotypes (e.g., DRw6) have proved to be genetically heterogeneous, sequence-based analysis has revealed specific alleles that are more highly associated with a particular disease. For example, the DRw6 serotype has a relative risk of 2.5 for the autoimmune dermatologic disease, *Pemphigus vulgaris* (PV), whereas a particular $DQ\beta$ allele, $DQ\beta1.3$ (one of six $DQ\beta$ alleles on DRw6, DQw1 haplotypes) has a relative risk of approximately 100 (Scharf et al. 1988). Unlike serologic or restriction-fragment-length polymorphism analysis of class II polymorphism, PCR-based analysis reveals not only that two alleles are different, but also how and where they differ. In the example above, the susceptible $DQ\beta1.3$ allele differs from another nonsusceptible allele by only a valine to aspartic acid substitution at position 57, implicating this residence as a critical element in susceptibility. (It should be noted that other non-susceptible $DQ\beta$ alleles also encode aspartic acid at position 57, so that it is the allele rather than an isolated residue that confers susceptibility.) A more detailed discussion of the relationship of class II sequence polymorphism, as detected by PCR, and disease susceptibility is presented in Horn et al. (1988), Scharf et al. (1988), and Todd et al. (1987). In general, specific combinations of class II alleles are most highly associated with autoimmune disease (e.g., IDDM, celiac disease [CD], and juvenile rheumatoid arthritis [JRA]).

By comparing the sequences of susceptible and nonsusceptible alleles, the amino acid at position 57 of the $DQ\beta$ chain and the amino acids around position 70 of the $DR\beta$I and $DP\beta$ chains (Bugawan et al. 1988b) have been revealed as critical residues in susceptibility to IDDM, PV, CD (Bugawan et al. 1989), and JRA (A. Begovich et al., in prep.). Moreover, the evolutionary analysis of the $DQ\beta$ polymorphism in nonhuman primates (see above) has revealed a balanced polymorphism between Asp-57 and alanine, valine, or serine at this position in all primate class II β-chains. The maintenance of this polymorphism suggests, as do the disease susceptibility studies, that this residue is functionally important. As noted above, it is the ability of PCR to amplify a specific segment of genomic DNA that has made possible the analyses of the HLA class II polymorphism described here.

REFERENCES

Bugawan, T.L., R.K. Saiki, C.H. Levenson, R.M. Watson, and H.A. Erlich. 1988a. The use of non-radioactive oligonucleotide probes to analyze enzymatically amplified DNA for prenatal diagnosis and forensic HLA typing. *Bio/Technology* **6**: 943.

Bugawan, T.L., G. Angelini, J. Larrick, S. Auricchio, G.B. Ferrara, and H.A. Erlich. 1989. A combination of a particular HLA-DPβ allele and an HLA-DQ heterodimer confers susceptibility to coeliac disease. *Nature* (in press).

Bugawan, T.L., G.T. Horn, C.M. Long, E. Mickelson, J.A. Hansen, G.B. Ferrara, G. Angelini, and H.A. Erlich. 1988b. Analysis of HLA-DP allelic sequence polymorphism using the *in vitro* enzymatic DNA amplification of DPα and DPβ loci. *J. Immunol.* **141**: 4024.

Gyllensten, U.B. and H.A. Erlich. 1988. Generation of single-stranded DNA by the polymerase chain reaction and its application to direct sequencing of the HLA-DQα locus. *Proc. Natl. Acad. Sci.* **85**: 7652.

———. 1989. Ancient roots for polymorphism at the HLA-DQα locus in primates. *Proc. Natl. Acad. Sci.* (in press).

Horn, G.T., T.L. Bugawan, C. Long, and H.A. Erlich. 1988. Allelic sequence variation of the HLA-DQ loci: Relationship to serology and insulin-dependent diabetes susceptibility. *Proc. Natl. Acad. Sci.* **85**: 6012.

Mullis, K.B. and F. Faloona. 1987. Specific synthesis of DNA in vitro via a polymerase catalysed chain reaction. *Methods Enzymol.* **155**: 335.

Saiki, R., P.S. Walsh, C.H. Levenson, and H.A. Erlich. 1989. Genetic analysis of amplified DNA with immobilized sequence-specific oligonucleotide probes. *Proc. Natl. Acad. Sci.* (in press).

Saiki, R.K., C.-A. Chang, C.H. Levenson, T.C. Warren, C.D. Boehm, H.H. Kazazian, Jr., and H.A. Erlich. 1988a. Diagnosis of sickle cell anemia and β-thalassemia with enzymatically amplified DNA and non-radioactive allele-specific oligonucleotide probes. *N. Engl. J. Med.* **319**: 537.

Saiki, R., S. Scharf, F. Faloona, K. Mullis, G. Horn, H.A. Erlich, and N. Arnheim. 1985. Enzymatic amplification of β-globin genomic sequences and restriction site analysis for diagnosis of sickle cell anemia. *Science* **230**: 1350.

Saiki, R.K., D.H. Gelfand, S. Stoffel, S. Scharf, R.H. Higuchi, G.T. Horn, K.B. Mullis, and H.A. Erlich. 1988b. Primer-directed enzymatic amplification of DNA with a thermostable DNA polymerase. *Science* **239**: 487.

Scharf, S.J., A. Friedman, C. Brautbar, F. Szafer, L. Steinman, G. Horn, U. Gyllensten, and H.A. Erlich. 1988. HLA class II allelic variation and susceptibility to *Pemphigus vulgaris. Proc. Natl. Acad. Sci.* **85**: 3504.

Todd, J.A., J.I. Bell, and H.O. McDevitt. 1987. HLA-DQβ gene contributes to susceptibility and resitance to insulin-dependent diabetes mellitus. *Nature* **329**: 599.

Mapping Immune Response Genes as Autoimmune Disease Susceptibility Loci

J.A. Todd[1] and A.N. Roberts[2]

[1]Nuffield Department of Surgery and Institute of Molecular Medicine
John Radcliffe Hospital, Headington
Oxford OX3 9DU, United Kingdom

[2]MRC Immunochemistry Unit, Department of Biochemistry
University of Oxford, OX1 3QU, United Kingdom

The utility of the polymerase chain reaction (PCR) (Saiki et al. 1985) is strikingly demonstrated in the analysis of the highly polymorphic genes involved in the immune response. In particular, the major histocompatibility complex (MHC) on human chromosome 6, composed of over 3×10^6 bp and containing perhaps as many as 100 genes (many of which are polymorphic), provides a daunting task for molecular genetics. One of the major reasons that this part of the genome is so intensely researched is that it has been associated with susceptibility to over 40 human diseases, and yet precise localization of disease-associated *MHC* genes, until recently, has remained elusive. This is mainly because of the observed linkage disequilibrium between *MHC* genes, which has made linkage analysis within families uninformative because of infrequent recombination between candidate disease loci. Linkage disequilibrium is likely to be common within the genome and will hinder precise localization of disease-specific genes. PCR has allowed us to circumvent these problems by enabling rapid and accurate analysis of the structures of candidate genes within the area of linkage disequilibrium from large numbers of individuals. We describe these applications in the context of the mapping of the MHC class II genes, *HLA-DQ*, as a susceptibility loci for insulin-dependent (type I) diabetes mellitus (IDDM), and also other PCR methods we have used to clone and to analyze new genes, such as the gene encoding the type II diabetes-associated peptide, amylin.

PCR from RNA
By late 1985, serological studies and restriction-fragment-length polymorphism (RFLP) analysis had shown that at least

one disease-susceptibility gene for autoimmune IDDM lay within the *DR-DQ* subregions of the MHC (Kim et al. 1984). The most likely candidate was the *DQB1* locus, but it could not be ruled out that the "real" susceptibility gene was closely linked to *DQB1* because of the strong linkage disequilibrium between *DR* and *DQ* genes. At this time, it became apparent that the polymorphism present in the amino-terminal domains of class II molecules was likely to be functional in the interaction of these proteins with foreign antigens during an immune response (Mengle-Gaw and McDevitt 1985; Germain and Malissen 1986). IDDM results from an autoimmune destruction of pancreatic insulin-producing β-cells that depends on the activity of T lymphocytes. T-cell activation is determined by MHC polymorphism. Hence, class II genes are attractive candidate disease-susceptibility loci. We therefore used PCR to amplify class II gene segments from cDNA synthesized from RNA, prepared from peripheral blood lymphocytes (PBL), and sequenced these to see if any polymorphism (not detectable by either serology or RFLP) correlated with disease susceptibility (Todd et al. 1987). In contrast with classic cDNA library preparation, we were able to analyze several genes from six patients with IDDM and also from several patients with myasthenia gravis and multiple sclerosis in a relatively short time (Todd et al. 1988).

Total RNA (15 μg or less; poly[A] purification is not necessary), prepared from PBLs from about 50 ml of blood, is subjected to cDNA synthesis (oligo[dT] primed, using Life Sciences' reverse transcriptase; Gubler and Hoffman 1983), and then a fraction of the first strand product (10% or less; second strand synthesis is not required) is added to the PCR reaction, with the appropriate primers. Originally, we used Klenow enzyme, which has a much lower error rate than *Taq* polymerase, but its use is more expensive and time-consuming than *Taq*-catalyzed PCR (Saiki et al. 1988). PCR products are gel-purified and cloned into an M13 vector designed specifically to facilitate cloning of unphosphorylated blunt-ended molecules (Waye et al. 1985; P. Carter, unpubl.). For blunt-end cloning, we have found that Klenow end-filling of the *Taq*-generated PCR product increases the cloning efficiency by at least twofold. The problem of *Taq* (and Klenow) polymerase sequence errors is solved by sequencing several clones of each allele or, if necessary, carrying out duplicate PCR reactions for each sample. It takes 5–6 days to go from blood collection to sequence analysis. This time can be reduced by scaling down both the preparation of PBLs (Potter and Potter 1988) and the RNA preparation (e.g., Wilkinson 1988). Also, if the PCR product contains mainly one sequence, direct DNA sequencing can be employed.

The *MHC* class II region contains pseudogenes that can contaminate PCR products if genomic DNA is used. RNA amplification avoids this problem, and, in addition, we have designed class II primers that do not amplify these pseudogene sequences (J.A. Todd, unpubl.). In general, we recommend starting with cDNA for PCR analysis of coding sequences.

Oligonucleotide Dot-blot Analysis

Sequence analysis of class II alleles from patients and controls revealed that a particular *DQB1* polymorphism is significantly correlated with susceptibility to IDDM, thus suggesting that *DQ* does directly encode disease susceptibility (Todd et al. 1987; Horn et al. 1988). We were able to confirm this result by analysis of class II polymorphisms in a large number of patients and controls using allele-specific oligonucleotide (ASO) dot-blot analysis, that is the detection of single or multiple DNA sequence polymorphisms by hybridization with 17-mer oligonucleotide probes (Saiki et al. 1986; Todd et al. 1987; Morel et al. 1988). In Caucasian populations, an aspartic acid residue at position 57 of the $DQ\beta$ chain (*DQ* and other class II molecules are α and β heterodimers) correlates with resistance to IDDM, individuals lacking $DQ\beta$ Asp-57 are over 100 times more likely to develop IDDM than Asp-57-positive individuals (Morel et al. 1988). Clinical trials of possible immunotherapies for IDDM will require identification of susceptible individuals in the population before the irreversible autoimmune destruction of the insulin-producing cells of the body begins. Large numbers of individuals can be screened by ASO analysis, and, although very few *DQ*-susceptible individuals (about 1 in 50) develop IDDM, the presence of protective *DQ* alleles should allow exclusion of Asp-57-positive subjects from clinical trials.

The development of *Taq* PCR (Saiki et al. 1988) makes the large scale analysis of DNA polymorphisms in populations and in families for linkage studies feasible. We have found that the quality of genomic DNA used in the PCR is important, and we employ a rapid and simple method (Graham 1978; A. Bushell, unpubl.) to prepare DNA from 10 ml of blood.

The requirement for unambiguous signals from PCR dot-

135

blot analysis cannot be overstated, particularly if DNA samples are from several different laboratories and may be degraded or contaminated with inhibitors of the reaction. We now routinely employ a "double PCR" approach to produce high-quality dot-blots with no significant variation between samples. This involves a primary PCR (30 cycles) with one set of primers, and then 2% of this product is subjected to a second PCR (30 cycles) using one of the original primers with a third "nested" primer placed 3' to one of the original primers.

This allows amplification of target DNA to levels that are easily cloned from very small amounts of starting material and when conditions are not optimal. PCR parameters that are titrated for each pair of oligonucleotides (17- to 20-mers, 50% GC content) include the Mg^{++} concentration and the annealing temperature, as recommended by Cetus (Saiki et al. 1988).

The lack of recombination because of linkage disequilibrium can also be resolved by analysis of haplotypes in different racial groups using these PCR methods. Race-specific haplotypes (e.g., globin and MHC haplotypes; Wainscoat et al. 1986; Fletcher et al. 1988) exist in populations of African descent. If a particular DNA polymorphism is situated in or very close to a disease allele, then the polymorphism should be associated with disease no matter on what haplotypic background the allele is. The identification of *DQB1* as a susceptibility locus for IDDM is further supported by the demonstration of a constant association of certain *DQB1* alleles with disease in a comparison of several Black and Caucasian haplotypes. Furthermore, *DQA1* polymorphism has also been shown to correlate with disease susceptibility (Todd et al. 1989). These methods are currently being used to study T-cell antigen-receptor gene expression qualitatively and quantitatively (Acha-Orbea et al. 1988; Roth et al. 1988).

Other PCR Applications: Cloning of the Amylin Gene

A common feature of the pancreases of type II diabetics is the deposition of amyloid. The major component of pancreatic amyloid is a peptide of 37 amino acids (Cooper et al. 1987). The in vitro biological activity of the peptide (amylin) suggests that it may be responsible for some of the characteristics of the pathogenesis of type II diabetes (Leighton and Cooper 1988). Screening of conventional cDNA and genomic libraries with oligonucleotide probes based on partial amino acid sequence can be technically demanding and time-consuming. An alternative is to use the oligonucleotides as primers in the PCR. Data from the cloning of homologous *MHC* genes suggested that efficient and specific PCR amplification could be achieved despite mismatches between the target and the primers. Two non-degenerate, "guessmer" oligonucleotide primers, biased for human codon usage (Lathe 1985), were synthesized based on the amino- and carboxy-terminal portions of the protein sequence of amylin. These oligonucleotides were used in a PCR with cDNA synthesized from RNA purified from human islets of Langerhans. A single band of the correct size was obtained (110 bp; annealing temperature 37°C), gel purified, and sequenced. The sequence corresponded to amylin, and the PCR product was used to probe a human genomic library. Phage clones containing the amylin gene were obtained and analyzed, revealing that both oligonucleotides had 17% mismatch with the amylin gene sequence (Cooper et al. 1989; A.N. Roberts, unpubl.). A similar approach using degenerate, "mixed" oligonucleotide PCR primers has been described recently (Knoth et al. 1988; Lee et al. 1988).

This is a powerful approach for the rapid cloning of new genes and genes that contain regions of sequence similarity (e.g., a homeo box domain). The approach will be facilitated when an efficient method for the generation of a priming site at the 5' end of cDNA is available, so that only one oligonucleotide sequence internal to the gene would be needed to obtain new sequences. We have had some success with the addition of linkers to cDNA (L. Timmerman et al., unpubl.). Application of the "inverted" PCR method (Triglia et al. 1988) to cDNA, to date, has not been productive using blunt-ended total cDNA as a substrate. An alternative approach (Loh et al. 1989) is to homopolymer tail cDNA (e.g., with dGTP) and to use poly(C) as a PCR primer.

REFERENCES

Acha-Orbea, H., D.J. Mitchell, L. Timmerman, D.C. Wraith, G.S. Tausch, M.K. Waldor, S.S. Zamvil, H.O. McDevitt, and L. Steinman. 1988. Limited heterogeneity of T cell receptors from lymphocytes mediating autoimmune encephalomyelitis allows specific immune intervention. *Cell* **54:** 263.

Cooper, G.J.S., A.C. Willis, A. Clark, R.C. Turner, R.B. Sim, and K.B.M. Reid. 1987. Purification and characterisation of a peptide from amyloid-rich pancreases of type 2 diabetic patients. *Proc. Natl. Acad. Sci.* **84:** 8628.

Cooper, G.J.S., A.N. Roberts, J.A. Todd, R. Sutton, A.J. Day, A.C. Willis, and B. Leighton. 1989. Amylin and non-insulin-dependent (type II) diabetes mellitus. In *Proceedings of the 13th International Diabetes Foundation Congress* (ed. R. Larkins et al.). Elsevier, Amsterdam. (In press.)

Fletcher, J., C. Mijovic, O. Odugbesan, D. Jenkins, A.R. Bradwell, and A.H. Barnett. 1988. Trans-racial studies implicate HLA-DQ as a component of genetic susceptibility to type I (insulin-dependent) diabetes. *Diabetologia* 31: 864.

Germain, R.N. and B. Malissen. 1986. Analysis of the expression and function of class II MHC-encoded molecules by DNA-mediated transfer. *Annu. Rev. Immunol.* 4: 281.

Graham, D. 1978. The isolation of high molecular weight DNA from whole organisms or large tissues. *Anal. Biochem.* 85: 609.

Gubler, U. and B.J. Hoffman. 1983. A simple and very efficient method for generating cDNA libraries. *Gene* 25: 263.

Horn, G.T., T.L. Bugawan, C.M. Long, and H.A. Erlich. 1988. Allelic sequence variation of the HLA-DQ loci: Relationship to serology and to insulin-dependent diabetes susceptibility. *Proc. Natl. Acad. Sci.* 85: 6012.

Kim, S.J., S.L. Holbeck, B. Nisperos, J.A. Hansen, H. Maeda, and G. Nepom. 1985. Identification of a polymorphic variant associated with HLA-DQw3 and characterised by specific restriction sites within the DQβ-chain gene. *Proc. Natl. Acad. Sci.* 82: 8139.

Knoth, K., S. Roberds, C. Poteet, and M. Tamkun. 1988. Highly degenerate, inosine-containing primers specifically amplify rare cDNA using the polymerase chain reaction. *Nucleic Acids Res.* 16: 10932.

Lathe, R. 1985. Synthetic oligonucleotide probes deduced from amino acid sequence data. *J. Mol. Biol.* 183: 1

Lee, C.C., X. Wu, R.A. Gibbs, R.G. Cook, D.M. Muzny, and C.T. Caskey. 1988. Generation of cDNA probes directed by amino acid sequence: Cloning of urate oxidase. *Science* 239: 1288.

Leighton, B. and G.J.S. Cooper. 1988. Pancreatic amylin and calcitonin gene-related peptide cause resistance to insulin in skeletal muscle *in vitro. Nature* 335: 632.

Loh, E.Y., J.F. Elliott, S. Cwirla, L.L. Lanier, and M.M. Davis. 1989. Polymerase chain reaction with single-sided specificity: Analysis of T cell receptor δ chain. *Science* 243: 217.

Mengle-Gaw, L. and H.O. McDevitt. 1985. Genetics and expression of mouse Ia antigens. *Annu. Rev. Immunol.* 3: 367.

Morel, P.A., J.S. Dorman, J.A. Todd, H.O. McDevitt, and M. Trucco. 1988. Aspartic acid at position 57 of the HLA-DQβ chain protects against type 1 diabetes: A family study. *Proc. Natl. Acad. Sci.* 85: 8111.

Potter, C.G. and A.C. Potter. 1988. A rapid and ultra-simplified method for separating lymphocytes from blood. *J. Immunol. Methods* 112: 143.

Roth, M.E., M.J. Lacy, L.K. McNeil, and D.M. Kranz. 1988. Selection of variable-joining region combinations in the α chain of the T cell receptor. *Science* 241: 1354.

Saiki, R.K., T.L. Bugawan, G.T. Horn, K.B. Mullis, and H.A. Erlich. 1986. Analysis of enzymatically amplified β-globin and HLA-DQα DNA with allele-specific oligonucleotide probes. *Nature* 324: 163.

Saiki, R.K., S. Scharf, F. Faloona, K.B. Mullis, G.T. Horn, H.A. Erlich, and N. Arnheim. 1985. Enzymatic amplification of β-globin genomic sequences and restriction site analysis for diagnosis of sickle cell anemia. *Science* **230**: 1350.

Saiki, R.K., D.H. Gelfand, S. Stoffel, S.J. Scharf, R. Higuchi, G.T. Horn, K.B. Mullis, and H.A. Erlich. 1988. Primer-directed enzymatic amplification of DNA with a thermostable DNA polymerase. *Science* **239**: 487.

Todd, J.A., J.I. Bell, and H.O. McDevitt. 1987. HLA-DQβ gene contributes to susceptibility and resistance to insulin-dependent diabetes mellitus. *Nature* **329**: 599.

Todd, J.A., C. Mijovic, J. Fletcher, D. Jenkins, A.R. Bradwell, and A.H. Barnett. 1989. Trans-racial mapping implicates both the HLA-DQA1 and DQB1 genes as susceptibility loci for insulin-dependent diabetes mellitus. *Nature* (in press).

Todd, J.A., H. Acha-Orbea, J.I. Bell, N. Chao, Z. Fronek, C.O. Jacob, M. McDermot, A.A. Sinha, L. Timmerman, L. Steinman, and H.O. McDevitt. 1988. A molecular basis for MHC class II-associated autoimmunity. *Science* **239**: 1026.

Triglia, A., M.G. Peterson, and D.J. Kemp. 1988. A procedure for in vitro amplification of DNA segments that lie outside the boundaries of known sequences. *Nucleic Acids Res.* **16**: 8186.

Wainscoat, J.S., A.V.S. Hill, A.L. Boyse, J. Flint, M. Hernandez, S.L. Thein, J.M. Old, J.R. Lynch, A.G. Falusi, D.J. Weatherall, and J.B. Clegg. 1986. Evolutionary relationships of human populations from an analysis of nuclear DNA polymorphisms. *Nature* **319**: 491.

Waye, M.M.Y., M.E. Verhoeyen, P.T. Jones, and G. Winter. 1985. EcoK selection vectors for shotgun cloning into M13 and deletional mutagenesis. *Nucleic Acids Res.* **13**: 8561.

Wilkinson, M. 1988. RNA isolation: A mini-prep method. *Nucleic Acids Res.* **16**: 10933

Length Polymorphisms in $(dC-dA)_n \cdot (dG-dT)_n$ Sequences Detected Using the Polymerase Chain Reaction

J.L. Weber

Marshfield Medical Research Foundation
Marshfield, Wisconsin 54449

One of the most abundant human interspersed repetitive DNA families are simple $(dC-dA)_n \cdot (dG-dT)_n$ ($[CA]_n$) dinucleotide repeats (Hamada and Kakunaga 1982; Sun et al. 1984). There are 50,000–100,000 blocks of $(CA)_n$ repeats in the genome with 15–30 repeats per block. We recently showed that lengths of blocks of these repeats are polymorphic within the human population (Weber and May 1988, 1989). $(CA)_n$ repeats therefore represent an extremely abundant new source of genetic markers.

At the start of this work, it was anticipated that $(CA)_n$ alleles might differ in size by as little as one repeat unit (2 bp). Standard restriction enzyme digestion, electrophoresis, blotting, and hybridization were judged to be impractical for routine detection of such slight size differences. Fortunately, the polymerase chain reaction (PCR) using thermostable *Taq* polymerase (Saiki et al. 1988) is much easier to apply to the typing of the $(CA)_n$ markers. Primers complementary to unique sequences flanking a particular block of $(CA)_n$ repeats are used to amplify and to label a small segment of DNA containing the repeats. The labeled DNA is then resolved by electrophoresis on polyacrylamide DNA sequencing gels and detected by autoradiography.

An example of this approach is shown in Figure 1. Three human genomic DNA samples (1, 2, and 3) served as templates for the amplification of the APOA2 $(CA)_n$ marker fragment. Labeling of the amplified DNA was performed by either incorporating $[\alpha\text{-}^{32}P]dATP$ into the interiors of both strands (leftmost set of three lanes) or by end labeling either the GT strand primer (middle three lanes) or the CA strand primer (rightmost three lanes) with $[\gamma\text{-}^{32}P]ATP$ and polynucleotide kinase prior to

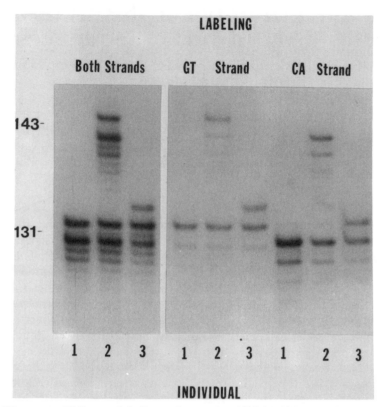

Figure 1 Different labeling schemes for APOA2 $(CA)_n$ marker fragments. Sizes in bases of the amplified fragments are indicated in the left margin.

amplification. The GT and CA strands migrate through the denaturing gels with slightly different mobilities. Each allele (when labeling with $[\alpha\text{-}^{32}P]dATP$) therefore results in two major bands on the autoradiographs. The lower (CA) band is more intense because it contains more adenines than the upper (GT) band. Individual 1 in Figure 1 is homozygous for a small allele of the APOA2 $(CA)_n$ marker; individual 2 is heterozygous for the small allele and another allele 12 bp larger; and individual 3 is heterozygous for two alleles that differ by only 2 bp.

To date, 16 human $(CA)_n$ repeat fragments have been amplified using the PCR, and all have proven to be polymorphic. The number of alleles found for these markers have ranged from 4 to 11, and the polymorphism information content values have ranged from 0.31 to 0.79 with an average value of

142

0.55. The $(CA)_n$ markers are therefore generally more informative than standard two allele restriction-fragment-length polymorphisms (Schumm et al. 1988) but are less informative than the most highly polymorphic minisatellites (Jeffreys et al. 1988). Several of the $(CA)_n$ markers have been typed in three-generation families, and all have shown standard Mendelian codominant inheritance. New $(CA)_n$ markers can be developed relatively easily by selecting genomic clones through hybridization to poly(dC-dA)·poly(dG-dT), sequencing the region containing the block of CA repeats, and then synthesizing PCR primers. Sequencing is simplified by cloning size-selected fragments of about 250 bp directly into M13.

Methodology for Marker Analysis

Standard PCRs (Saiki et al. 1985, 1988) for typing the $(CA)_n$ markers were carried out in 25 µl volumes containing 10–20 ng of genomic DNA template, 100 ng each oligodeoxynucleotide primer, 200 µM each dGTP, dCTP, and dTTP, 2.5 µM dATP, 1–2 µCi of [α-^{32}P]dATP at 800 Ci/mmoles or α-^{35}S-labeled dATP at 500 Ci/mmoles, 50 mM KCl, 10 mM Tris (pH 8.3), 1.5 mM $MgCl_2$, 0.01% gelatin, and about 0.5 unit of *Taq* polymerase (Cetus). Generally, samples were processed through 25 temperature cycles consisting of 1 minute at 94°C, 2 minutes at 55°C, and 1 minute at 72°C. The last elongation step was lengthened to 10 minutes. Aliquots of the amplified DNA were mixed with two volumes of formamide sample buffer and electrophoresed on standard denaturing polyacrylamide DNA sequencing gels. Gels were then dried, and the film was exposed. Optimal exposure times were about 2 days. The PCR conditions described above are close to standard with the exception of the 2.5 µM [dATP]. The concentration of this nucleotide was reduced to increase the specific activity of the labeled amplified DNA. Reducing the [dATP] also tends to diminish background fragments larger in size than the $(CA)_n$ fragments.

Efficiency in typing the $(CA)_n$ markers is improved by amplifying several fragments simultaneously in the same tube and also by resolving the amplified markers from several different $(CA)_n$ blocks in the same gel line. To date, up to four different marker fragments have been successfully amplified together, although background on the polyacrylamide gels tends to increase as more primers are added to the reaction. Similarly, since the largest and smallest alleles for most of the $(CA)_n$ marker fragments differ by less than 20 bp, it will be possible to analyze at least five or six markers on one gel lane. A single 60-lane sequencing gel can be used to determine 300 or more genotypes.

Artifacts of the Amplification Reactions

In Figure 1, a number of fainter bands are seen for each allele in addition to the two most intense bands. These extra bands are smaller in size than the most intense bands and differ in size from the major bands by one to several bases. Because the extra bands can make reading of the genotypes difficult, their elimination is desirable.

After excluding impurities in the primer preparations as a cause of the additional bands, two additional alternative explanations were considered: somatic mosaicism and PCR artifacts. Three different results rule out the presence of substantial somatic mosaicism. First, the additional bands are observed when cloned plasmid DNA is used as template; second, when template DNA from a heterozygous individual is reduced in the PCR to the point that heterozygosity is lost (only one of the two alleles are amplified), the extra bands are still present; and third, when DNA from 32 lymphocyte clones from each of two individuals (Nicklas et al. 1987) was used as the template, no allelic differences were detected between the clones.

Griffin et al. (1988) recently demonstrated that DNA fragments amplified by the PCR could not be efficiently ligated to blunt-ended vectors without first repairing the ends of the fragments with T4 DNA polymerase. To test whether "ragged" ends were responsible for the extra bands associated with $(CA)_n$-amplified fragments, amplified DNA was treated with the Klenow fragment of DNA polymerase I or with T4 DNA polymerase, as shown in Figure 2. In this experiment, aliquots of two different $(CA)_n$-amplified fragments, untreated from the PCR (C), were brought up to 200 μM [dATP] and then incubated at 37°C for 30 minutes with 6 units of Klenow enzyme (K), 1 unit of T4 DNA polymerase (P), or with no enzyme (T).

Both Klenow enzyme and T4 DNA polymerase simplified the banding pattern somewhat by eliminating extra bands that differed in size from the most intense bands by one base. These enzymes also reduced the size of the most intense bands by one base. The most likely explanation for these results is that the *Taq* polymerase produces a mixture of fragments during the PCR with different types of ends. The most intense bands in

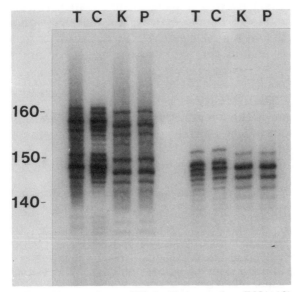

Figure 2 Treatment of amplified $(CA)_n$ marker DNA (*C*) with the Klenow fragment of DNA polymerase I (*K*), T4 DNA polymerase (*P*), or without added enzyme (*T*).

the untreated samples are likely derived from double-stranded molecules with single-base noncomplementary 3′ overhangs. The fainter bands, which are one base smaller than the major bands, are likely to be blunt ended. The 3′ – 5′ exonuclease activity of Klenow or T4 DNA polymerase converts the molecules with overhangs into blunt-ended molecules. Clark (1988) recently showed that a variety of DNA polymerases, including *Taq* polymerase, can add a noncomplementary extra base to the 3′ ends of blunt-ended molecules.

Remaining after Klenow or T4 DNA polymerase treatment are extra bands that differ in size from the major bands by multiples of two bases. Our unproven working hypothesis is that this subgroup of extra bands is due to the skipping of repeats by the *Taq* polymerase during the elongation cycles. Note that end-labeling PCR primers, as shown in Figure 1, leads only to the extra bands differing in size by multiples of two bases. It may be that a phosphate at the 5′ end of the primer inhibits the addition of the extra noncomplementary 3′ base.

In summary, the PCR is for practical purposes indispensible in the detection of $(CA)_n$ repeat length polymorphisms.

PCR permits genotypes to be determined efficiently using very small amounts of template DNA. Polymorphic human $(CA)_n$ markers should play a major role in the study of genetic disease genes and in the improvement of genetic and physical maps.

ACKNOWLEDGMENTS

I thank Jan Nicklas for the lymphocyte clones and Paula May for expert technical assistance. This work was supported by the Marshfield Clinic.

REFERENCES
Clark, J.M. 1988. Novel non-templated nucleotide addition reactions catalyzed by procaryotic and eucaryotic DNA polymerases. *Nucleic Acids Res.* **16:** 9677.
Griffin, L.D., G.R. MacGregor, D.M. Muzny, J. Harter, R.G. Cook, and E.R.B. McCabe. 1988. Synthesis of hexokinase 1 (HK1) cDNA probes by mixed oligonucleotide primed amplification of cDNA (MOPAC) using primer mixtures of high complexity. *Am. J. Hum. Genet.* (suppl.) **43:** A185.
Hamada, H. and T. Kakunaga. 1982. Potential Z-DNA forming sequences are highly dispersed in the human genome. *Nature* **298:** 396.
Jeffreys, A.J., N.J. Royle, V. Wilson, and Z. Wong. 1988. Spontaneous mutation rates to new length alleles at tandem-repetitive hypervariable loci in human DNA. *Nature* **332:** 278.
Nicklas, J.A., T.C. Hunter, L.M. Sullivan, J.K. Berman, J.P. O'Neill, and R.J. Albertini. 1987. Molecular analyses of *in vivo* hprt mutations in human T-lymphocytes. *Mutagenesis* **2:** 341.
Saiki, R.K., S. Scharf, F. Faloona, K.B. Mullis, G.T. Horn, H.A. Erlich, and N. Arnheim. 1985. Enzymatic amplification of β-globin genomic sequences and restriction site analysis for diagnosis of sickle cell anemia. *Science* **230:** 1350.
Saiki, R.K., D.H. Gelfand, S. Stoffel, S.J. Scharf, R. Higuchi, G.T. Horn, K.B. Mullis, and H.A. Erlich. 1988. Primer-directed enzymatic amplification of DNA with a thermostable DNA polymerase. *Science* **239:** 487.
Schumm, J.W., R.G. Knowlton, J.C. Braman, D.F. Barker, D. Botstein, G. Akots, V.A. Brown, T.C. Gravius, C. Helms, K. Hsiao, K. Rediker, J.G. Thurston, and H. Donis-Keller. 1988. Identification of more than 500 RFLPs by screening random genomic clones. *Am. J. Hum. Genet.* **42:** 143.
Sun, L., K.E. Paulson, C.W. Schmid, L. Kadyk, and L. Leinwand. 1984. Non-Alu family interspersed repeats in human DNA and their transcriptional activity. *Nucleic Acids Res.* **12:** 2669.
Weber, J.L. and P.E. May. 1988. An abundant new class of human DNA polymorphisms. *Am. J. Hum. Genet.* (suppl.) **43:** A161.
———. 1989. Abundant class of human DNA polymorphisms which can be typed using the polymerase chain reaction. *Am. J. Hum. Genet.* **44:** 388.

The Application of Polymerase Chain Reaction in Forensic Science

G.F. Sensabaugh and C. von Beroldingen

Forensic Science Group, Department of Biomedical and
Environmental Health Sciences, School of Public Health
University of California, Berkeley, California 94720

Biological evidence in forensic cases comes in many forms: blood stains on any imaginable surface, semen mixed with vaginal fluids, hairs, bits of bone, and tissue from under fingernails. A major objective in the analysis of such evidence is to develop information as to the identity of the individual who left the evidence behind. This has been traditionally approached by testing for blood group and protein genetic markers present in the evidence samples (Sensabaugh 1982). The advance of DNA technology over the past decade now provides a new approach, the analysis of genetic variation at the DNA level. This approach offers the promise of virtual absolute identification of source, a promise that has generated considerable excitement both in the forensic community and in the public media.

The Value of PCR Amplification in the Analysis of Evidence

The capacity of an evidence sample to be analyzed depends on the quantity and quality of DNA in the sample. The polymerase chain reaction (PCR) offers two significant advantages in this regard: (1) Amplification is possible from very small amounts of starting DNA. This allows the extraction of genetic information from evidence samples that contain too little cellular material for other genetic typing approaches (e.g., single shed hairs [Higuchi et al. 1988]). (2) Amplification is possible from degraded DNA (Paabo et al. 1988). Moderate to severe DNA strand breakage is found in a significant proportion of evidence samples. From an operational standpoint, PCR offers two additional advantages: It is relatively simple to perform, and results can be obtained within a short time frame, usually within 24 hours (Kazazian and Dowling 1988).

A current limitation on the usefulness of PCR in evidence

analysis is the existence of only one well-characterized typing system, the HLA-DQα system (Saiki et al. 1986; Erlich, this volume). DQα typing will differentiate two random individuals about 91% of the time. Under development are typing systems for the detection of other polymorphisms, including variable number tandem repeat polymorphisms; these will greatly extend the power of PCR-based genetic typing. Ultimately, PCR might be used to sequence highly informative regions of the genome, such as the mitochondrial D-loop region (Cann et al. 1987; Orrego et al. 1988; Paabo et al. 1988); there is some sentiment in the forensic community that DNA sequence profiling may be the approach of the future.

Fidelity of Amplification in Native and Damaged DNA

One of the major legal concerns about the application of PCR to evidence analysis is that the DNA being typed is a copy, not the original; it must be clear beyond a reasonable doubt that PCR gives true genetic types. The intrinsic fidelity of *Taq* polymerase has been estimated in independent measurements to be on the order of 1×10^{-4} to 2×10^{-4} misincorporation per nucleotide (Saiki et al. 1988; Tindall and Kunkel 1988). It can be shown both by calculation and by experiment that this misincorporation rate should not lead to typing error.

The proportion of PCR product sequences containing any specific misincorporation depends on the initial copy number of the template DNA and on the cycle in which the misincorporation first occurs. The worst case situation would occur if the misincorporation event was at the first cycle with an initial copy number of 1 (i.e., a single sperm cell). In this situation, 25% of the products would contain the misincorporation. When the initial template copy number is larger or when the misincorporation occurs at a later cycle, the proportion of product sequences containing any particular error is correspondingly decreased. Under practical testing conditions, the calculations show the proportion of product sequences containing any specific misincorporation is not high enough to affect typing results significantly.

We have also put the misincorporation question to an experimental test, using a portion of the human hemoglobin βA (Hbβ^A) chain sequence as the test system. This sequence contains a *Dde*I restriction site that overlaps the site of mutation that differentiates the Hbβ^A sequence from the Hbβ^S sequence. This Hbβ^A sequence was amplified from human genomic DNA through 30 cycles, interrupting the process every 5 cycles to restrict the accumulated product with *Dde*I; the amplification process was then continued through another 20 cycles. This test scheme thus selects for misincorporations occurring at the *Dde*I site. Of the 12 possible misincorporations at this site, 1 gives rise to the Hbβ^S sequence; however, no significant Hbβ^S sequence was detected in the PCR product in this experiment. Thus, even under conditions designed to generate misincorporation typing error, none could be produced.

In the forensic context, the question of amplification fidelity extends to include DNA templates that have suffered damage (e.g., Friedberg 1985; Imlay and Linn 1988). Does *Taq* polymerase read through damage sites, and if so, does it insert the correct nucleotide? We have found that *Taq* polymerase stops short at damage sites created by UV radiation; this is typical of the behavior of DNA polymerases at damage sites (e.g., Clark and Beardsley 1987). Thus, the available evidence suggests that DNA damage poses little risk of introducing misincorporations that could lead to typing error. We are continuing our studies on this matter.

Contamination

Contamination of evidence samples, reagents, and working solutions by PCR products generated in the laboratory poses another significant forensic concern. This is really a quality control issue, and the lessons of dealing with contamination in microbiology and cell culture laboratories are pertinent. It is clear that a one-way line of flow from sample preparation to PCR to the genetic typing test must be set up; this should keep PCR products away for the early stages of analysis where contamination can play havoc.

ACKNOWLEDGMENTS
This work was supported by National Institute of Justice grant 86-IJ-CX-0044. We are grateful to Russell Higuchi, Henry Erlich, and Edward Blake for helpful discussions.

REFERENCES
Cann, R.L., M. Stoneking, and A.C. Wilson. 1987. Mitochondrial DNA and human evolution. *Nature* **325**: 31.
Clark, J.M. and G.P. Beardsley. 1987. Functional effects of cis-thymine glycol lesions on DNA synthesis in vitro. *Biochemistry* **26**: 5398.
Friedberg, E.C. 1985. DNA damage. In *DNA repair*, p.1. W.H. Freeman, New York.

Higuchi, R., C.H. von Beroldingen, G.F. Sensabaugh, and H.A. Erlich. 1988. DNA typing from single hairs. *Nature* **332:** 543.

Imlay, J.A. and S. Linn. 1988. DNA damage and oxygen radical toxicity. *Science* **240:** 1302.

Kazazian, H.H. and C.E. Dowling. 1988. Laboratory implications of automated polymerase chain reaction. *Am. Biotechnol. Lab.* **6(6):** 23.

Orrego, C., A.C. Wilson, and M.C. King. 1988. Identification of maternally related individuals by amplification and direct sequencing of a highly polymorphic, noncoding region of mitochondrial DNA. *Am. J. Hum. Genet.* **43:** A219.

Paabo, S., J.A. Gifford, and A.C. Wilson. 1988. Mitochondrial DNA sequences from a 7000-year old brain. *Nucleic Acids Res.* **16:** 9775.

Saiki, R.K., T.L. Bugawan, G.T. Horn, K.B. Mullis, and H.A. Erlich. 1986. Analysis of enzymatically amplified β-globin and HLA-DQᾶ with allele specific oligonucleotide probes. *Nature* **324:** 163.

Saiki, R.K., D.H. Gelfand, S. Stoffel, S.J. Scharf, R. Higuchi, G.T. Horn, K.B. Mullis, and H.A. Erlich. 1988. Primer directed enzymatic amplification of DNA with a thermostable DNA polymerase. *Science* **239:** 487.

Sensabaugh, G.F. 1982. Biochemical markers on individuality. In *Forensic science handbook* (ed. R. Saferstein), p. 338. Prentice Hall, Englewood Cliffs, New Jersey.

Tindall, K.R. and T.A. Kunkel. 1988. Fidelity of DNA synthesis by *Thermus aquaticus* DNA polymerase. *Biochemistry* **27:** 6008.

Application of the Polymerase Chain Reaction to the Detection of Human Retroviruses

S. Kwok, D.H. Mack, D.N. Kellogg, N. Mckinney, F. Faloona, and J.J. Sninsky

Department of Infections Diseases, Cetus Corporation
Emeryville, California 94608

The application of the polymerase chain reaction (PCR) (Saiki et al. 1985; Mullis and Faloona 1987) to the study of human retroviruses has presented the procedure with several unique challenges. First, sensitive detection requires the amplification of an extremely small number of viral copies present in highly complex nucleic acid mixtures. Early studies demonstrated as few as 1 transcriptionally active infected cell per 10,000 uninfected cells in the lymph nodes of infected individuals (Harper et al. 1986). Second, the detection of only one member of a family of closely related but distinct elements is required. Not only are there multiple members of exogenous virus families (human immunodeficiency virus [HIV] types 1 and 2; human T-cell leukemia/lymphoma virus [HTLV] types I and II), but the human genome is riddled with multiple families of endogenous retroviral sequences, some containing large numbers of members. Third, the well-documented viral heterogeneity within and between individuals harboring these pathogens requires the targeting of conserved regions for amplification to allow the detection of historic, contemporary, and future viral variants. Fourth, the human retroviruses characterized to date and those yet to be identified are associated with, responsible for, or implicated in an ever-increasing number of human diseases and disorders. Despite numerous valiant efforts, the first human retrovirus was identified less than 10 years ago (Poiesz et al. 1980). The use of viral cultures to detect new members of the known human retroviral families is hampered by the necessity to propagate the target cell and the cumbersome and time-consuming nature of this procedure. Procedures to facilitate the detection and characterization of related but distinct members of retroviruses would greatly accelerate research in this area. Fifth, the generation by PCR of vast quantities of DNA fragments of defined sequence can lead to contamination or carry-over of the products of one reaction to subsequent PCR reactions prior to amplification.

Sensitive Detection

Retroviral infection is likely to represent a dynamic equilibrium between infected and uninfected cells, transcriptionally active and dormant infected cells, as well as infected cells in peripheral blood and those sequestered in privileged sites in the body. The persistence of human retroviruses for long periods of time prior to the development of a disease, even in the context of a substantial immune response, is an earmark of these infections. The unusually low number of infected cells is presumably a consequence of the transcriptional dormancy of infected cells. Both viral regulatory genes that suppress transcription and translation, as well as the inability for retroviral promoters to function efficiently in nondividing cells, play a role in the attenuation of viral replication. Another complicating factor in retroviral detection is the relatively small number of proviral copies in infected cells. The use of PCR in initial studies on HIV suggested the procedure would contribute to the sensitive detection of infected cells (Kwok et al. 1987, 1989; Ou et al. 1988). Although HIV RNA can be targeted for amplification by PCR (Byrne et al. 1988; Hart et al. 1988; Murakawa et al. 1988), to date, the use of HIV DNA as a template appears more sensitive. The increased specificity of PCR provided by *Taq* polymerase (Saiki et al. 1988), the judicious selection of primers for amplification, and optimization of the various reaction components have resulted in the reproducible detection of 10–20 copies of proviral DNA in the equivalent of 150,000 cells (Kwok et al. 1989). Given the observed viral heterogeneity (Hahn et al. 1986; Saag et al. 1988; Goodenow et al. 1989), efforts to ensure the detection of most viral variants requires continued evaluation of the efficiency of the chosen primers and probes to amplify and detect divergent sequences. This level of analytical sensitivity provides for the detection of most infected individuals. In addition, although not performed in a rigorously quantitative manner, the signal intensity of the amplified product suggests that the level of infected cells in either symptomatic or asymptomatic individuals may be 10- to 100-fold greater than was appreciated earlier.

Endogenous Retroviral Families

The retroviral life-cycle includes the synthesis of a complementary, double-stranded, circular DNA provirus that covalently integrates into the host's cellular genome. The proviral genome remains integrated for the life of the cell. The infection of germ-line cells results in the eventual vertical transmission of the viral sequence to the progeny of the infected host. Past studies of these endogenous viral sequences have led to several conclusions (see Table 1). To date, it appears as though the retroviral subfamily Oncovirinae have participated in the evolution of mammalian and avian cellular genomes to a greater extent than either the Lentivirinae or the hepadnaviruses (a class of viruses that are similar to retroviruses, that replicate through an RNA intermediate using a reverse transcriptase, but that package DNA rather than RNA), but additional studies are required to resolve this issue. Since thousands of copies of these elements are present, the amplification of these sequences must be considered when conserved regions of exogenous retroviruses are targeted. Primers for the long terminal repeat (LTR) and the region of the retroviral genome that is complementary to tRNA (the initiator of plus strand replication) were recently designed for HTLV-I. Unfortunately, amplification using a standard set of conditions and HTLV-I-infected cells led to a weak signal by Southern blot analysis. Examination of an ethidium-bromide-stained gel indicated that numerous DNA fragments of varying molecular weights were generated (Fig. 1, lane 1). We postulated that perhaps endogenous viral sequences were being amplified, thereby resulting in a decreased specific PCR product. Decreases in the enzyme, magnesium chloride, and primer concentrations led to a dramatic increase in the specificity and sensitivity of the

Table 1 Endogenous Retroviral Elements of Mammalian Genomes

Multiple families reside in the genome

Family members vary in sequence and structure

Nearly intact and substantially altered members are present

Numerous, nontandem copies are present

Resulted from infection after speciation

May share significant sequence homology with pathogenic exogenous retroviruses

Role in pathogenesis is unclear

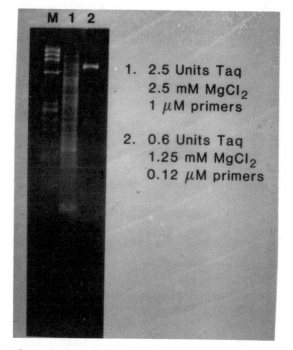

Figure 1 Optimization of PCR conditions to circumvent amplification of endogenous viral sequences related to HTLV-I in the human genome.

reaction (Fig. 1, lane 2). Optimal detection of exogenous retroviral sequences therefore requires not only highly efficient primers but also those capable of amplifying all viral variants as well as ones that, under the selected conditions, do not amplify remnants of past retroviral infections. Correspondingly, a more complete characterization of endogenous elements will result from the application of PCR (Shih et al. 1989). Of particular note is the presence in feline genomic DNA of a small number of endogenous proviral copies that are highly homologous to the exogenous feline leukemia virus except for a divergent U3 region of the LTR (Casey et al. 1981). One should therefore not exclude the possibility that there exist exogenous human retroviruses similar in sequence to elements within the human genome.

Identification of New Retroviruses
The human retroviruses characterized to date will represent only a fraction of the viruses responsible for human diseases.

Since related viruses share short hyphenated regions of homology, PCR is ideally suited to facilitate the detection of uncharacterized viruses. Initially, the judicious selection of highly conserved regions of the *trans*-activator gene of the HTLV family allowed the simultaneous detection of both the type I and type II viruses (Kwok et al. 1988a). In addition, we recently employed short, "degenerate" oligonucleotides as primers for a region of the reverse transcriptase gene of the hepadnaviruses to direct the coincident amplification of the mammalian (hepatitis B virus, woodchuck hepatitis virus, and ground squirrel hepatitis virus) and avian (duck hepatitis B virus) members of this viral family (Mack and Sninsky 1988). We expect that the conserved regions of the genes responsible for replication and regulation of viral life cycle will serve as productive targets for PCR. Furthermore, the development of generic PCR assays that amplify conserved regions of related viruses has significant diagnostic potential for cost-effective screening.

Carry-over

Although extraneous DNA from multiple sources may serve as a template for amplification, thereby leading to false positive reactions, the predominant source appears to be the PCR product from previous PCR reactions. Therefore, caution should be used when numerous amplifications of the same primer pair system are carried out (Schochetman et al. 1988). The implications of false positives in the context of HIV-1 diagnostic screening prompted us to develop and recommend a series of precautions that should be incorporated to reduce the frequency and amount of carry-over (Table 2). The conscientious following of these recommendations should dramatically decrease the likelihood of carry-over.

CONCLUSION

Despite these challenges, the study of human retroviruses has and will continue to benefit from the application of PCR. PCR has demonstrated clinical utility for detecting infection prior to the generation of detectable antibodies (Imagawa et al. 1989), for neonatal screening (Laurie et al. 1988; M.F. Rogers et al.; E.G. Chadwick et al.; both unpubl.), resolution of individuals

Table 2 Precautions to Minimize PCR Carry-over (Contamination)

Physically separate pre- and post-PCR reactions
Use positive displacement pipets
Aliquot reagents
Modify lab behavior (radioactivity analogy)

Judiciously select the number and type of controls
    small number of low-copy positive controls
    large number of reagent controls
    large number of negative controls

Replicate results
Use highly characterized samples

with ambiguous or indeterminate serological status (B. Jackson et al., unpubl.), and virus typing for individuals infected with either or both types of HIV (Rayfield et al. 1988) and HTLV (Kwok et al. 1988a,b). The differential diagnosis, particularly for HTLV-I versus HTLV-II, and possibly for HIV-1 and HIV-2, will assist individual follow-up. The delineation of the pathogenesis, if any, of HTLV-II and the human spumavirus (Maurer et al. 1988) merits intense investigation. It should be noted, however, that when PCR is used to link the characterized human retroviruses to diseases, one must take into consideration not only the carry-over issue but also supportive epidemiological and serological data. We expect the role that endogenous retroviral families play in eukaryotic development and perhaps pathogenesis will begin to be deciphered with PCR. Finally, PCR may prove fruitful in the search for new members of characterized viral families.

ACKNOWLEDGMENTS

The authors extend their sincere appreciation to the other members of the Cetus PCR group for ongoing criticism and support. In addition, these studies could not have been carried out without the productive and congenial collaborations with Drs. B. Poiesz, G. Schochetman, B. Jackson, S. Wain-Hobson, and their colleagues, as well as the members of the Multicenter AIDS Cohort Study.

REFERENCES

Byrne, B.C., J.J. Li, J.J. Sninsky, and B.J. Poiesz. 1988. Detection of HIV-1 RNA sequences by in vitro DNA amplification. *Nucleic Acids Res.* **16:** 4165.

Casey, J.W., A. Roach, J.I. Mullins, K. Bauman-Burck, M.O. Nicolson, M.B. Gardner, and N. Davidson. 1981. the U3 portion of feline leukemia virus DNA identifies horizontally acquired proviruses in leukemic cats. *Proc. Natl. Acad. Sci.* **78**: 7778.

Goodenow, M., T. Huet, W. Saurin, S. Kwok, J.J. Sninsky, and S. Wain-Hobson. 1989. HIV-1 isolates are rapidly evolving quasispecies: Evidence for viral mixtures and preferred nucleotide substitutions. *J. Acq. Immunodef. Dis.* (in press).

Hahn, B.H., G.M. Shaw, M.E. Taylor, R.R. Redfield, P.D. Markham, S.Z. Saldhuddin, F. Wong-Staal, R.C. Gallo, E.S. Parks, and W.P. Parks. 1986. Genetic variation in HTLV-III/LAV over time in patients with AIDS or at risk for AIDS. *Science* **232**: 1548.

Harper, M.H., L.M. Marselle, R.C. Gallo, and F. Wong-Staal. 1986. Detection of lymphocytes expressing human T-lymphotrophic virus type III in lymph nodes and peripheral blood from infected individuals by in situ hybridization. *Proc. Natl. Acad. Sci.* **83**: 772.

Hart, C., G. Schochetman, T. Spira, A. Lifson, J. Moore, J. Galphin, J.J. Sninsky, and C.-Y. Ou. 1988. Direct detection of HIV RNA expression in seropositive subjects. *Lancet* **II**: 596.

Imagawa, D.T., M.H. Lee, S.M. Wolinsky, K. Sano, F. Morales, S. Kwok, J.J. Sninsky, P.G. Nishanian, J. Giorgi, J.L. Fahey, J. Dudley, B.R. Visscher, and R. Detels. 1989. Human immunodeficiency virus type 1 infection in homosexual men who remain seronegative for prolonged periods. *N. Engl. J. Med.* **320**: 1458.

Kwok, S., G. Ehrlich, B. Poiesz, R. Kalish, and J.J. Sninsky. 1988a. Enzymatic amplification of HTLV-I viral sequences from peripheral blood mononuclear cells and infected tissues. *Blood* **72**: 1117.

Kwok, S., D. Kellogg, G. Ehrlich, B. Poiesz, S. Bhagavati, and J.J. Sninsky. 1988b. Characterization of a sequence of human T cell leukemia virus type I from a patient with chronic progressive myelopathy. *J. Infect. Dis.* **158**: 1193.

Kwok, S., D.H. Mack, K.B. Mullis, B.J. Poiesz, G. Ehrlich, D. Blair, A. Friedman-Kien, and J.J. Sninsky. 1987. Identification of human immunodeficiency virus sequences by using in vitro enzymatic amplification and oligomer cleavage detection. *J. Virol.* **61**: 1690.

Kwok, S., D.H. Mack, J.J. Sninsky, G.D. Ehrlich, B. Poiesz, N. Dock, H.J. Alter, D. Mildvan, and M.H. Grieco. 1989. Diagnosis of human immunodeficiency virus in seropositive individuals: Viral sequences in peripheral blood mononuclear cells. In *HIV detection by genetic engineering methods* (ed. P.A. Luciw and K.S. Steimer), p. 243. Marcel Dekker, New York.

Laurie, F., V. Courgnaud, C. Rouzioux, S. Blanche, F. Veber, M. Burgard, C. Jacomet, C. Griscelli, and C. Brechot. 1988. Detection of HIV DNA in infants and children by means of the polymerase chain reaction. *Lancet* **II**: 538.

Mack, D.H. and J.J. Sninsky. 1988. A sensitive method for the identification of uncharacterized viruses related to known virus groups: Hepadnavirus model system. *Proc. Natl. Acad. Sci.* **85**: 6977.

Maurer, B., H. Bannert, G. Darai, and R.M. Flugel. 1988. Analysis of the primary structure of the long terminal repeat and the *gag* and *pol* genes of the human spumaretrovirus. *J. Virol.* **62**: 1590.

Mullis, K.B. and F.A. Faloona. 1987. Specific synthesis of DNA in vitro via a polymerase catalyzed chain reaction. *Methods Enzymol.* **155:** 35.

Murakawa, G.J., J.A. Zaia, P.A. Spallone, D.A. Stephens, B.E. Kaplan, R.B. Wallace, and J.J. Rossi. 1988. Direct detection of HIV-1 RNA from AIDS and ARC patients samples. *DNA* **7:** 287.

Ou, C.-Y., S. Kwok, S.W. Mitchell, D.H. Mack, J.J. Sninsky, J.W. Krebs, P. Feorino, D. Warfield, and G. Schochetman. 1988. DNA amplification for direct detection of HIV-1 in DNA of peripheral blood mononuclear cells. *Science* **239:** 295.

Poiesz, B.J., F.W. Ruscetti, A.F. Gazdar, P.A. Bunn, J.D. Minna, and R.C. Gallo. 1980. Detection and isolation of type C retrovirus particles from fresh and cultured lymphocytes of a patient with cutaneous T cell lymphoma. *Proc. Natl. Acad. Sci.* **77:** 7415.

Rayfield, M., K. De Cock, W. Heyward, L. Goldstein, J. Krebs, S. Kwok, S. Lee, J. McCormick, J.M. Moreau, K. Odehouri, G. Schochetman, J.J. Sninsky, and C.-Y. Ou. 1988. Mixed human immunodeficiency virus (HIV) infection in an individual: Demonstration of both HIV type 1 and type 2 proviral sequences by using polymerase chain reaction. *J. Infect. Dis.* **158:** 1170.

Saag, M.S., B.H. Hahn, J. Gibbons, Y. Li, E.S. Parks, W.P. Parks, and G.M. Shaw. 1988. Extensive variation of human immunodeficiency virus type 1 in vivo. *Science* **334:** 440.

Saiki, R.K., S.J. Scharf, F.A. Faloona, K.B. Mullis, G.T. Horn, H.A. Erlich, and N. Arnheim. 1985. Enzymatic amplification of β-globin genomic sequences and restriction site analysis for the diagnosis of sickle cell anemia. *Science* **230:** 1350.

Saiki, R.K., D.H. Gelfand, S. Stoffel, S.J. Scharf, R. Higuchi, G.T. Horn, K.B. Mullis, and H.A. Erlich. 1988. Primer-directed enzymatic amplification of DNA with a thermostable DNA polymerase. *Science* **239:** 487.

Schochetman, G., C.-Y. Ou, and W.K. Jones. 1988. Polymerase chain reaction. *J. Infect. Dis.* **158:** 1154.

Shih, A., R. Misra, and M.G. Rush. 1989. Detection of multiple, novel reverse transcriptase coding sequences in human nucleic acids: Relation to primate retroviruses. *J. Virol.* **63:** 64.

Use of Polymerase Chain Reaction in the Detection, Quantification, and Characterization of Human Retroviruses

B.J. Poiesz,[1] G.D. Ehrlich,[1] B.C. Byrne,[1] M. Abbott,[1] S. Kwok,[2] and J. Sninsky[2]

[1]Division of Hematology/Oncology, Health Science Center
State University of New York, Syracuse, New York 13210
[2]Cetus Corporation, Emeryville, California 94608

The Retroviruses

Retroviruses are the etiologic agents of a host of diseases found in vertebrates. These include malignancies such as lymphomas, leukemias, sarcomas, and carcinomas; autoimmune diseases such as arthritis and lupus; and cytopathic diseases leading to anemias and immunodeficiency states. There are four well-characterized human retroviruses. Human T-cell lymphoma/leukemia virus types I and II (HTLV-I and HTLV-II) are oncornaviruses. HTLV-I is believed to be the etiologic agent of adult T-cell lymphoma/leukemia and a progressive neurological disorder formerly called tropical spastic paraparesis, now termed HTLV-I-associated myelopathy (Bhagavati et al. 1988; Erlich and Poiesz 1988). HTLV-I is also associated with immunodeficiency. HTLV-I infection is endemic in southern Japan, the Caribbean, and Central Africa. In the United States, HTLV-I infection is most prevalent in the southeastern section of the country among rural blacks, but it has also been identified in patients throughout America and appears to be increasing in prevalence among intravenous drug abusers (Ratner and Poiesz 1988). HTLV-II is approximately 65% homologous to HTLV-I. It has been associated with rare cases of T-cell and hairy cell leukemias, but its exact disease association and endemic area of infection are unclear at this time. HTLV-II infection is also being diagnosed with greater frequency in American intravenous drug abusers. HTLV-V is a recently described retrovirus with limited homology with HTLV-I. It has been identified in a few patients with CD4+ cutaneous T-cell lymphoma, but the virus has not been completely characterized, and

159

its disease association is poorly understood. Human immuno-deficiency virus types-1 and -2 (HIV-1 and HIV-2) are the etiologic agents of AIDS. HIV-2 infection is primarily confined to western Africa, whereas HIV-1 is responsible for the current pandemic of AIDS.

Problems Associated with Diagnosis

Most humans infected with a retrovirus manifest no overt symptoms for many years following infection. Some, but not all, eventually evolve through protracted prodromal stages of their disease prior to the development of a terminal illness. The therapeutic options for patients with retrovirally induced terminal illnesses are few, and none to date are curative (Poiesz et al. 1988). Hence, considerable effort has been made to identify infected asymptomatic patients to prevent spread of the infection to others and specify the exact nature of their illnesses (Duggan et al. 1988) or to initiate earlier antiviral treatment strategies. Considerable use has been made of serological assays to detect infected individuals. Such techniques have served society well as screening assays to protect recipients of donated blood products. However, because of the latency associated with retroviral infections, the possibility of seronegative, infected states exists. Furthermore, for a number of reasons (e.g., determining prognosis, evaluating treatment efficacy, and/or analyzing genetic variation of different virus isolates), it is important to have the capability to test for the presence of the retroviruses themselves. Accordingly, we have adapted the polymerase chain reaction (PCR) technique for the detection, quantification, and characterization of human retroviruses (Kwok et al. 1987, 1988b,c; Karpatkin et al. 1988).

Sensitive and Specific Detection by PCR

Primer pairs have been developed that allow for the sensitive and specific detection of HIV-1, HIV-2, HTLV-I, and HTLV-II proviral DNAs using either the Klenow fragment of *Escherichia coli* DNA polymerase I or, more recently, *Taq* DNA polymerase to support enzymatic amplification (Fig. 1). Similarly, these techniques have been employed for the detection of retroviral RNA after first producing cDNA, using cloned Moloney murine leukemia virus reverse transcriptase and then allowing the standard *Taq*-polymerase-based system to amplify the cDNA (Byrne et al. 1988). Prior treatment of the sample to be amplified with DNase allows for the determination of levels of

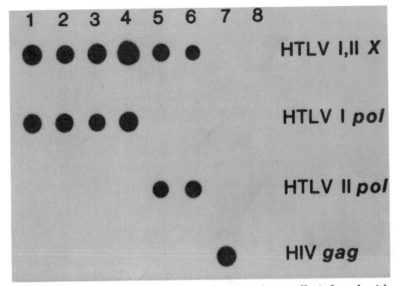

Figure 1 Analysis of PCR-amplified DNA from cells infected with various human retroviruses. PCR amplification was performed using 30 cycles and *Taq* polymerase. The lanes represent DNA extracted from cell lines containing integrated HTLV-I (lanes *1–4*), HTLV-II (lanes *5* and *6*), HIV-1 (lane *7*), and an uninfected T-cell line, Molt-4 (lane *8*). Row *1* represents PCR amplification that used primers SK43/44 and probe SK45, which contain sequences conserved between HTLV-I and HTLV-II; row *2* represents primers SK54/55 and SK56 that are HTLV-I specific; row *3* represents primers SK58/59 and SK60 that are HTLV-II specific; and row *4* represents primers SK38/39 and SK19 that are HIV-1 specific.

RNA expression in the specimen. Detection systems utilized have included direct incorporation of radioactive isotope into the amplified product; or hybridization techniques employing slot-blot analysis, liquid hybridization, or oligomer restriction detection formats (Abbott et al. 1988). Direct incorporation of radioisotopes works for certain retroviral primer pairs; however, for most primer pairs we have chosen to perform ultimately a hybridization detection step because of the often observed nonspecific amplification of presumably endogenous retroviral sequences located within the human genome. Oligomer restriction detection, because of its requirement for the generation of a specific endonuclease restriction site, created by the hybridization of the detector sequence to the amplified target, has proven to be perhaps too specific as a

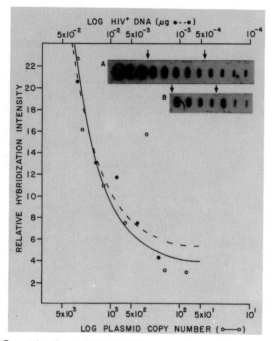

Figure 2 Quantitative analysis of the reaction products of amplified HIV-1-positive DNA diluted into negative DNA. Decreasing concentrations of HIV-1-positive DNA samples were amplified in a total of 1 μg of DNA per reaction mixture. Amplification was accomplished using the HIV-1 *gag*-specific primer-detector system (SK38/39, SK19). The amplified DNA was slot blotted and hybridized with the ^{32}P-labeled detector. Curves were fitted by linear regression using only the densitometric scans resulting from those dilutions within the arrows on the insets; the coefficient of coincidence: $r^2 = 0.94$ for both plasmid and virus sets. (Inset *A*) Dilutions of DNA isolated from the chronically infected cell line HUT78/ HIV$_{AAV}$. The film demonstrates a quantitative decrease in hybridization intensity as target HIV-positive DNA is diluted from 200 ng per reaction mixture to 100 pg. (Inset *B*) Dilutions of DNA from a plasmid carrying a 2.2-kb insert encompassing the amplified region. The film depicts hybridization resulting from amplification of dilutions from 3000 to 80 copies of HIV DNA per reaction mixture.

screening technique for retroviruses. We have instead used a relatively less stringent slot-blot analysis to screen samples for positivity and a more stringent liquid hybridization technique and/or sequence analyses to evaluate the specific product.

Quantification of the amplified product can be achieved by comparing a laser-generated densitometric scan of the autoradiographic signal of a patient's sample with a serially diluted standard control (Fig. 2). Similar analysis can be performed for human genes (e.g., β-globin), and results can be expressed as copies of retroviral DNA per unit of human β-globin amplified. Ultimate characterization of amplified product has been conducted via cloning of the amplified products into M13 and subsequently performing sequence analyses (Kwok et al. 1988a).

To date, results indicate that PCR detection is a highly sensitive format. Using 1 µg of fresh peripheral blood mononuclear cell DNA (≈150,000 cells) and 30 cycles of amplification and at least two viral specific primer pairs, our detection rates for HIV-1 and HTLV-I in seropositive individuals have approximated 80% and 100%, respectively. We believe that the relatively lower rates of HIV-1 detection are secondary to lower in vivo copy numbers and to sequence variability between the various HIV isolates versus the HTLV-I isolates. Indeed, when the DNA from 10^6 cells of HIV-1-infected individuals is analyzed, the sensitivity of the PCR system increases significantly.

One important quality of PCR is its ability to distinguish between HTLV-I and HTLV-II infections. In a recent survey of 169 intravenous drug abusers in New York City, 17 (10%) persons tested positive for antibodies to both HTLV-I and HTLV-II. Analysis by PCR indicates that 10 (6%) and 13 (7%) subjects were positive for HTLV-I and HTLV-II sequences. Perhaps of even greater interest is the fact that a total of 11 (7%) subjects were found to be seronegative but PCR positive for either HTLV-I or HTLV-II.

The use of PCR theoretically could facilitate the detection of variant retroviruses closely or distantly related to the known human retroviruses. To that end, we have recently screened HTLV-I-seronegative patients with CD4+ lymphomas for the presence of HTLV-I- or HTLV-II-positive sequences. Several patients have been found to score positive consistently with some, but not all, HTLV-I-specific primer pairs. The data suggest the possibility of the presence of either defective or novel retroviral infections in these patients.

In summary, PCR has been adapted for the facile, quantitative, and specific amplification of retroviral nucleic acids. the clinical use of this technology should prove invaluable in the detection, counseling, and treatment of infected individuals. PCR should also allow for relatively rapid characterization of retroviral variants and offers theoretical promise for the discovery of new retroviruses and/or retroviral disease associations in man.

ACKNOWLEDGMENTS
We gratefully acknowledge the contributions of Dr. Jian Jun Li, Janice Andrews, Virginia Bryz-Gornia, and Lynn Zaumetzer for PCR analyses, and Mary Rubert for the preparation of the figures.

REFERENCES
Abbott, M., B. Poiesz, J. Sninsky, S. Kwok, B. Byrne, and G. Ehrlich. 1988. A comparison of methods for the detection and quantification of the polymerase chain reaction. *J. Infect. Dis.* **158:** 1158.

Bhagavati, S., G. Ehrlich, R. Kula, S. Kwok, J. Sninsky, V. Udani, and B. Poiesz. 1988. Detection of human T-cell lymphoma/leukemia virus-type I (HTLV-I) in the spinal fluid and blood of cases of chronic progressive myelopathy and a clinical, radiological and electrophysiological profile of HTLV-I associated myelopathy. *N. Engl. J. Med.* **318:** 1141.

Byrne, B., J. Li, J. Sninsky, and B. Poiesz. 1988. Detection of HIV-1 RNA sequences by in vitro DNA amplification. *Nucleic Acids Res.* **16:** 4165.

Duggan, D., G. Ehrlich, F. Davey, S. Kwok, J. Sninsky, J. Goldberg, L. Baltrucki, and B. Poiesz. 1988. HTLV-I induced lymphoma mimicking Hodgkin's disease: Diagnosis by polymerase chain reaction amplification of specific HTLV-I sequences in tumor DNA. *Blood* **71:** 1027.

Ehrlich, G. and B. Poiesz. 1988. Clinical and molecular parameters of HTLV-I infection. *Clin. Lab. Med.* **8:** 65.

Karpatkin, S., M. Nardi, E. Lennette, B. Byrne, and B. Poiesz. 1988. Anti-HIV-1 antibody complexes on platelets of seropositive thrombocytopenic homosexuals and narcotic addicts. *Proc. Natl. Acad. Sci.* **85:** 9763.

Kwok, S., G. Ehrlich, B. Poiesz, S. Bhagavati, and J. Sninsky. 1988a. Characterization of a HTLV-I sequence from a patient with chronic progressive myelopathy. *J. Infect. Dis.* **158:** 1193.

Kwok, S., G. Ehrlich, B. Poiesz, R. Kalish, and J. Sninsky. 1988b. Enzymatic amplification of HTLV-I viral sequences from peripheral blood mononuclear cells and infected tissues. *Blood* **72:** 1117.

Kwok, S., D. Mack, K. Mullis, B. Poiesz, G. Ehrlich, D. Blair, A. Friedman-Kien, and J. Sninsky. 1987. Identification of HIV viral sequences using in vitro enzymatic amplification and oligomer cleavage detection. *J. Virol.* **61:** 1690.

Kwok, S., D. Mack, G. Ehrlich, B. Poiesz, N. Dock, H. Alter, D. Mildvan, M. Grieco, and J. Sninsky. 1988c. Diagnosis of human immunodeficiency virus in seropositive individuals: Enzymatic amplification of HIV viral sequences in peripheral blood mononuclear cells. In *Genetic engineering approaches to AIDS diagnosis* (eds. P.A. Luciw and K.S. Steimer), p. 241. Marcel Dekker, New York.

Poiesz, B., G. Ehrlich, L. Papsidero, and J. Sninsky. 1988. Detection of human retroviruses. In *AIDS: Etiology, diagnosis, treatment and prevention* (eds. V. DeVita et al.), p. 137. Lippincott, Philadelphia.

Ratner, L. and B. Poiesz. 1988. Human T-cell lymphotropic virus type I associated leukemias in a non-endemic region. *Medicine* **67:** 401.

Polymerase Chain Reaction in AIDS Research

C.-Y. Ou and G. Schochetman

AIDS Program, Center for Infectious Diseases
Centers for Disease Control, Atlanta, Georgia 30333

The identification of persons with exposure to human immunodeficiency virus (HIV) is indicated by the presence of antibodies to HIV, viral antigens, and/or viruses in the bodily components. In the last few years, approaches employing nucleic-acid-based technology have been widely used to detect HIV genetic information. However, because of the low number of infected lymphocytes in a seropositive person, conventional techniques, such as Southern blot analysis, were not sensitive enough to detect and to characterize HIV genetic information. Polymerase chain reaction (PCR) was first used to detect HIV proviral DNA in HIV coculture (Kwok et al. 1987) and was then extended to detect directly HIV proviral DNA in the peripheral blood mononuclear cells (PBMCs) of infected persons (Ou et al. 1988). PCR has since been employed in many studies as a novel molecular tool to amplify HIV DNA for detection and detailed characterization (Schochetman et al. 1988). In this paper, we review our use of PCR in AIDS research.

RESULTS AND DISCUSSION

Although serologic assays identify persons with previous exposure to HIV, they do not specifically indicate current infection. Laboratory evidence for current infection requires virus culture and/or a positive detection of viral antigens. Virus culture takes at least 3–4 weeks and lacks sensitivity in that viruses cannot be consistently isolated from infected persons. Antigen assays are based on the capture of HIV antigens with polyclonal or monoclonal antibodies that are immobilized on a solid support. The major drawback of the assay is that the HIV antigens in most of the clinical specimens are complexed to viral-specific antibodies preventing antigen detection. Free p24 viral antigen usually appears late in the course of immunodeficiency syndrome when there is antigen excess in the circulation. In some cases, free antigens can be detected in the first few months of infection prior to the development of humoral antibodies. Thus, the value of an antigen test is limited.

PCR overcomes the limitations of virus culture and antigen testing and represents a revolutionary technology for the detection and characterization of HIV. Current procedures for the detection of HIV DNA in the blood of infected persons with PCR involves four components: (1) isolation of PBMCs from blood using Ficoll-hypaque, (2) preparation of DNA, (3) amplification of short DNA segments of HIV using specific primer pairs, and (4) detection of amplified DNA by hybridization. The entire procedure takes 2–3 days. Viral RNA can also be detected by converting it to DNA using reverse transcriptase (RTase). The resultant DNA can then be amplified by PCR (Byrne et al. 1988; Hart et al. 1988; Murakawa et al. 1988).

HIV-1 and HIV-2 share considerable sequence homology and thus elicit considerable serologic cross-reactivity. The presence of unique sequences allows the design of specific primer pairs to be used as site-directed molecular probes to differentiate these two viruses. The first West African patient residing in the United States with an HIV-2 infection had an antibody response specific to HIV-2 but not to HIV-1. To confirm that this person was infected with HIV-2, PCR was performed with DNA from the patient's PBMCs, using HIV-1- and HIV-2-specific primers and probes. HIV-2 but not HIV-1 sequences were detected, indicating HIV-2 infection only (Weiss et al. 1988). This was also confirmed by isolation of HIV-2 from this person.

In a study of HIV-1 and HIV-2 prevalence in high risk groups in the Ivory Coast, we (Rayfield et al. 1988) identified a person coinfected with HIV-1 and HIV-2. The serum of this person reacted to both viruses, using whole virus lysate enzyme immunoassays, Western blot analysis, and site-directed serology using synthetic peptides specific for a region in the transmembrane protein (gp41). HIV-1- and HIV-2-specific primers, representing sequences from the long terminal repeat and *gag* regions, were used to establish the coexistence of both viruses in this person's lymphocytes. A recombinant virus could arise if there is coinfection of a single lymphocyte with both viruses. Our initial seroprevalence study in the Ivory Coast indicates that 14% of the seropositive persons were seroreactive to both viruses. The true prevalence of HIV-1 and HIV-2 in this population may have to be ultimately resolved by highly sensitive PCR and site-directed serologic tests.

The clinical value of PCR is well demonstrated when it is applied to perinatal diagnosis of HIV infection. Infants born to seropositive mothers have HIV antibodies of maternal origin because of cross-placental transfer. Because the maternal antibodies can last for 6–15 months after birth, a positive serologic test on the infants does not necessarily indicate infection. Early identification of HIV-infected infants is important so that therapeutic intervention can be implemented in the first stages of the infection. It was reported that 6 of 14 PBMCs collected 2–3 days after the infants were born to seropositive mothers were found to be HIV DNA positive by PCR (Laure et al. 1988). These infants were symptom-free and only one of them had detectable viral antigens in the serum. The proportion of PCR positivity agrees with the finding that the risk of transmission of HIV during pregnancy is 30–40%. However, there is no additional laboratory or clinical evidence to support infection in these PCR-positive infants. Prospective follow-up of these infants is necessary to substantiate the PCR results. The AIDS Program of the Centers for Disease Control, New York City Department of Health, and Montefiore Medical Center undertook a study to evaluate the diagnostic and prognostic value of PCR in infants (M.F. Rogers et al., unpubl.). Of 15 infants with PBMC collected within the neonatal period (first 28 days after birth), 6 were positive by PCR. Five of these six infants developed AIDS within the first year of life. In contrast, proviral sequences were only detected in neonatal specimens of one of eight infants who developed nonspecific symptoms of HIV infection.

Proviral DNA was detected in later specimens in another three infants. Eighteen infants who were clinically healthy for at least 15 months and who seroreverted are all PCR negative. Those in whom HIV DNA is detected early in life are likely to develop AIDS later.

The RTase of HIV is error-prone (Preston et al. 1988; Roberts et al. 1988) and is at least partially associated with the extensive genetic heterogeneity observed with HIV (Hahn et al. 1986; Starcich et al. 1986; Saag et al. 1988). Because of the low level of HIV in infected persons, sequence determination requires propagation and expansion in vitro to obtain sufficient virus to analyze. Because of the numerous virus replicative cycles during this expansion, the estimated evolutionary rate based on viruses derived from culture may be overestimated. PCR can be used to eliminate this problem.

We have employed a primer pair (C071, TGTGGAGGGGAA TTCTTCTACTGTAA, and CO72, TATAGAATTCACTTCTCCA ATTGTCCCTCAT) derived from two conserved sequences of the gp120 *env* gene. The intervening sequences include two variable domains, V4 and V5, and a constant domain, C3, which contains the CD4-binding domain. *Eco*RI restriction endonuclease sites (GAATTC) were incorporated into the primers so that the amplified products could be cloned into the *Eco*RI site of an M13 vector for sequencing. Individual phages containing the amplified HIV *env* segment were sequenced. Assuming the efficiency of priming and duplication of DNA strands in the amplification step are the same for each proviral DNA in a patient's PBMCs, the sequence profile should reflect the viral profile as present in the patient's PBMCs. Two PBMC specimens were collected at different time intervals from a homosexual man to evaluate the evolution of his HIVs in vivo over time. One specimen was obtained while the patient was seronegative, and the other specimen was collected 5 months later after he seroconverted. We first determined the type and rate of sequence changes that are contributed by the amplification and M13 cloning procedures by using an HIV DNA clone with a defined intervening sequence of 262 bp as a starting template. In 13 recombinant clones of a total of 3406 bases, 8 base substitutions were observed, and the calculated overall basal error rate was 0.23 per 100 bases. No deletions or insertions were observed.

Ten clones were sequenced from the preseroconversion specimen of the homosexual man, and five were shown to be identical. Each of the other four had only one base substitution at different positions in the 253-bp intervening sequences. The remaining clone had three substitutions. Since the basal error rate is 0.58 (0.23% x 253) substitution per clone, a clone with only one base substitution could be a result of the amplification and cloning process. It therefore appears that there was only a very small number of viral variants in this patient at the time of specimen collection. His postseroconversion specimen showed 6 identical clones out of 12 sequenced. The other six isolates possessed only one substitution each, all located at different positions. The six identical clones were the same as those found in the first collection. Since the number of substitutions per isolate was not much higher than the basal error rate, we cannot conclude that these minor variants exist in vivo.

In contrast, a specimen derived from a hemophilic man

revealed a different picture. Of 13 isolates sequenced, none were identical to each other. There were many base substitutions, insertions, and deletions. This result indicated that the hemophilic man who received contaminated blood factors might have been exposed to multiple viruses. It will be interesting to monitor HIV profiles from a variety of persons to see if one or a few variants may predominate in a person over time.

CONCLUSION
Detection of HIV DNA in bodily components of infected persons by PCR can be a useful adjunct to HIV serologic diagnosis. However, a PCR test requires sophisticated laboratory skills and facilities and is not currently suitable for routine use. Here, we have presented three studies to demonstrate the utilities of PCR to (1) differentiate HIV-1 and HIV-2, (2) to identify early infants who acquired HIV infection from seropositive mothers, and (3) to study the evolution of HIV in the lymphocytes of infected persons.

REFERENCES
Byrne, B.C., J.J. Li, J. Sninsky, and B.J. Poiesz. 1988. Detection of HIV-1 RNA sequences by an in vitro DNA amplification. *Nucleic Acids Res.* **16:** 4165.
Hahn, B.H., G.M. Shaw, M.E. Taylor, R.R. Redfield, P.D. Markham, S.Z. Salhuddin, F. Wong-Staal, R. Gallo, E.S. Parks, and W.P. Parks. 1986. Genetic variation in HTLV-III/LAV over time in patients with AIDS or at risk for AIDS. *Science* **232:** 1548.
Hart, C., G. Schochetman, T. Spira, A. Lifson, J. Moore, J. Galphin, J. Sninsky, and C.-Y. Ou. 1988. Direct detection of HIV RNA expression in seropositive subjects. *Lancet* **II:** 596.
Kwok, S., D.H. Mack, K. Mullins, B. Poiesz, G. Ehrlich, D. Blair, A. Friedman-Kien, and J. Sninsky. 1987. Identification of human immunodeficiency virus sequences by using in vitro enzymatic amplification and oligomer cleavage detection. *J. Virol.* **61:** 1690.
Laure, F., V. Courgnaud, C. Rouzioux, S. Blanche, F. Veber, M. Burgard, C. Jacomet, C. Griscelli, and C. Brechot. 1988. Detection of HIV DNA in infants and children by means of the polymerase chain reaction. *Lancet* **II:** 538.
Murakawa, G.J., J.A. Zaia, P.A. Spallone, D.A. Stephens, B.E. Kaplan, R.B. Wallace, and J.J. Rossi. 1988. Direction detection of HIV-1 RNA from AIDS and ARC patient samples. *DNA* **7:** 287.
Ou, C.-Y., S. Kwok, S.W. Mitchell, D.H. Mack, J. Sninsky, J.W. Krebs, P. Feorino, D. Warfield, and G. Schochetman. 1988. DNA amplification for direct detection of HIV-1 in DNA of peripheral blood mononuclear cells. *Science* **239:** 295.
Preston, B., B.J. Poiesz, and L.A. Loeb. 1988. Fidelity of HIV-1 reverse transcriptase. *Science* **242:** 1168.

Rayfield, M., K. DeCock, W. Heyward, L. Goldstein, J. Krebs, S. Kwok, S. Lee, J. McCormick, J.M. Moreau, K. Odehouri, G. Schochetman, J. Sninsky, and C.-Y. Ou. 1988. Mixed HIV infection of an individual: Demonstration of both HIV-1 and HIV-2 proviral sequences by polymerase chain reaction. *J. Infect. Dis.* **158:** 1170.

Roberts, J.D., K. Bebenek, and T.A. Kunkel. 1988. The accuracy of reverse transcriptase from HIV-1. *Science* **242:** 1167.

Saag, M.S., B.H. Hahn, J. Gibbons, Y. Li, E.S. Parks, W.P. Parks, and G.M. Shaw. 1988. Extensive variation of human immunodeficiency virus type-1 in vivo. *Science* **334:** 440

Schochetman, G., C.-Y. Ou, and W.K. Jones. 1988. Polymerase chain reaction. *J. Infect. Dis.* **158:** 1154.

Starcich, B.R., B.H. Hahn, G.M. Shaw, P.D. McNeely, S. Modrow, H. Wolf, E.S. Parks, W.P. Parks, S.F. Josephs, R.C. Gallo, and F. Wong-Staal. 1986. Identification and characterization of conserved and variable regions in the envelope gene of HTLV-III/LAV, the retrovirus of AIDS. *Cell* **45:** 637.

Weiss, S.H., J. Lombardo, J. Michaels, L.R. Sharer, M. Tayyarah, J. Leonard, A. Mangia, P. Kloser, S. Sathe, R. Kaplia, N.W. Williams, R. Altman, J. French, W.E. Perkins, Genetic Systems Corp., and AIDS Program, Center for Infectious Diseases, CDC. 1988. AIDS due to HIV-2 infection — New Jersey. *Morbidity Mortality Wkly. Rep.* **37:** 33.

Rapid Identification of mRNA Splice Sites by the Polymerase Chain Reaction

T.R. Broker, L.T. Chow, and M.O. Rotenberg[1]

Biochemistry Department, University of Rochester School of Medicine
Rochester, New York 14642

The polymerase chain reaction (PCR) (Saiki et al. 1985) has been used for a variety of purposes. In this paper, we describe the adaptation of this technique for the identification of splice sites in extremely rare mRNAs that have otherwise been refractory to conventional biochemical analyses on the nucleotide level.

INTRODUCTION

Human papillomavirus (HPV) types cause warts, dysplasias, and carcinomas of the anogenital and oral mucosa. They cannot be propagated in cultured cells. Viral mRNAs in human lesions are usually rare. Electron microscopy of RNA:DNA heteroduplexes formed between cloned HPV DNAs and mRNAs isolated from lesions has shown that families of alternatively spliced transcripts originate from several promoters and use one of two alternative polyadenylation sites (Chow et al. 1987a,b). So far, only one HPV cDNA has been cloned from human lesions (Fig. 1, species a) (Nasseri et al. 1987); it corresponds to the most abundant viral mRNA encoding an E1^E4 fusion protein. Comparison of genomic and cDNA sequences of several papillomaviruses with the structures of mRNAs mapped by electron microscopy has led to predictions of splice sites for several mRNAs of human and animal papillomaviruses (Chow et al. 1987a).

RESULTS

To ascertain the exact structures of the HPV type 11 mRNAs and the proteins encoded, we amplified cDNAs using the PCR

---

[1]Present address: Biophysics Department, University of Rochester School of Medicine, Rochester, New York 14642.

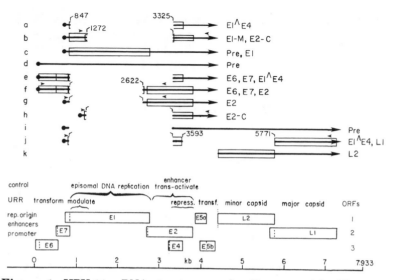

Figure 1 HPV-11 mRNA structures and coding potentials. The circular genome of HPV-11 is represented in a linear fashion. (*Top*) mRNA species are represented by arrows in the 5' to 3' direction with gaps signifying intron sequences spliced out of the mRNA. Primers used to generate the corresponding cDNA fragments are denoted by arrowheads above each mRNA pointing in the 5' to 3' direction. The mRNA splice sites determined are indicated only once, although some may have been sequenced (bent tick marks) from different cDNA species. The open reading frames (ORFs) in the mRNAs inferred from its structure and sequence are marked by open boxes and the protein(s) encoded are indicated to the right of each species. (*Bottom*) Functions associated with proteins derived from the ORFs in the genomic DNA sequence (Dartmann et al. 1986) are partially based on bovine papillomavirus type-1 (BPV-1) (for review, see Broker and Botchan 1986). The positions of putative initiation codons in the ORFs are marked by vertical lines.

technique. Total RNA was extracted from a condylomatous cyst developed from HPV-11-infected human foreskin implanted under the renal capsule of an athymic mouse (Kreider et al. 1987). Poly(A)⁺ RNA was primed with oligo(dT) for first-strand cDNA synthesis. The oligonucleotide primers for the PCR were designed on the basis of their abilities to prime the synthesis of cDNA copies of only one or a couple of the HPV-11 messages as predicted by RNA:DNA heteroduplex analysis (Fig. 1) (Chow et al. 1987a). Because of the overlapping nature of most of the mRNAs, each HPV-11 cDNA was separately amplified. The time permitted for primer extension during each PCR cycle was kept to a minimum to diminish the possibility of generating cDNA copies of unspliced HPV-11 RNA precursors. At the conclusion of a 30-cycle reaction, the mixture was extracted once with an equal volume of chloroform, and 1/20 of the recovered products was subjected to 30 additional cycles of amplification. Aliquots of the final PCR reaction products were fractionated via agarose gel electrophoresis followed by ethidium bromide staining. In each case, the size of the most prominent product corresponded to that predicted for the targeted viral mRNA species. The PCR products were extracted from the gel, and cDNA sequences spanning the splice sites were determined directly or after cloning into a plasmid vector. The sequencing primer was either one of the two PCR primers or an internal primer close to the suspected mRNA splice site. Comparison of the cDNA sequences with the published genomic DNA sequence established the splice sites in nearly all the spliced mRNAs that have been previously detected (Fig. 1, species b and h, f and g, and j). The likely coding capacities of each of these mRNAs were deduced. For instance, the E2-C protein is encoded by species h, deriving its initiation codon from the upstream exon; it functions as a repressor for enhancers in the upstream regulatory region (URR) (Chin et al. 1988). Species b could encode the E1-M protein, which modulates extrachromosomal viral DNA copy number in the BPV-1 system (Berg et al. 1986); it could also encode the E2-C protein if internal reinitiation during translation is possible. Species f and g each encode the enhancer regulating protein, E2 (Hirochika et al. 1987; M. Rotenberg et al., in prep.). Species f could also encode the E6 protein and perhaps the E7 protein as well. Interestingly, mRNA species j, expressed only in the most differentiated cells (Chow et al. 1987c), contains the entire coding region for the abundant E1^E4 protein (Doorbar et al. 1986; Breitburd et al. 1987; Brown et al. 1988), which is also encoded by the predominant species—an mRNA present in almost all the infected cells (Chow et al. 1987c). In addition, mRNA species j also contains the L1 ORF encoding the major capsid protein synthesized only in the most differentiated cells. These results are summarized in Figure 1.

DISCUSSION

The extremely high sensitivity of the PCR technique demonstrates that this approach can be used as an alternative to conventional methods for the isolation of cDNA copies of rare mRNAs. It is rapid, efficient, and precise, eliminating the need for the tedious screening of cDNA libraries. In combination with direct DNA sequencing, the PCR approach allows for the quick identification of splice junctions within amplified cDNA species. It should be noted, however, that the technique as used here is not useful for defining the natural 5' and 3' termini of corresponding mRNAs. For the explicit purpose of splice site determination, cDNA isolation via PCR requires at least some advance knowledge of the approximate structures of the mRNA species and the corresponding genomic sequences.

ACKNOWLEDGMENTS

This study was supported by U.S. Public Health Service grant CA-36200. We are grateful to Dr. John Kreider for the HPV-11 condylomatous cyst. We thank the Cetus Corporation for the generous gifts of the Perkin-Elmer Cetus thermal cycler and the oligonucleotide primers.

REFERENCES

Berg, L., M. Lusky, A. Stenlund, and M.R. Botchan. 1986. Repression of bovine papilloma virus replication is mediated by a virally encoded *trans*-acting factor. *Cell* 46: 753.

Breitburd, F., O. Croissant, and G. Orth. 1987. Expression of human papillomavirus type-1 E4 gene products in warts. *Cancer Cells* 5: 115.

Broker, T.R. and M. Botchan. 1986. Papillomaviruses: Retrospectives and prospectives. *Cancer Cells* 4: 17.

Brown, D.R., M.T. Chin, and D.G. Strike. 1988. Identification of human papillomavirus type 11 E4 gene products in human tissue implants from athymic mice. *Virology* 165: 262.

Chin, M.T., R. Hirochika, H. Hirochika, T.R. Broker, and L.T. Chow. 1988. Regulation of human papillomavirus type 11 enhancer and E6 promoter by activating and repressing proteins from the E2 open reading frame: Functional and biochemical studies. *J. Virol.* 62: 2994.

Chow, L.T., M. Nasseri, S.M. Wolinsky, and T.R. Broker. 1987a. Human papillomavirus types 6 and 11 mRNAs from genital condylomata. *J. Virol.* 61: 2581.

Chow, L.T., S.S. Reilly, T.R. Broker, and L.B. Taichman. 1987b. Identification and mapping of human papilloma virus type 1 mRNA recovered from plantar warts and infected epithelial cell culture. *J. Virol.* 61: 1913.

Chow, L.T., H. Hirochika, M. Nasseri, M.H. Stoler, S.M. Wolinsky, M.T. Chin, R. Hirochika, D.S. Arvan, and T.R. Broker. 1987c. Human papillomavirus gene expression. *Cancer Cells* 5: 55.

Dartmann, K., E. Schwarz, L. Gissmann, and H. zur Hausen. 1986. The nucleotide sequence and genome organization of human papilloma virus type 11. *Virology* 151: 124.

Doorbar, J., D. Campbell, R.J.A. Grand, and P.H. Gallimore. 1986. Identification of the human papilloma virus-1a E4 gene products. *EMBO J.* **5:** 355.

Hirochika, H., T.R. Broker, and L.T. Chow. 1987. Enhancers and *trans*-acting transcriptional factors of papillomaviruses. *J. Virol.* **61:** 2599.

Kreider, J.W., M.K. Howett, A.E. Leure-Dupree, R.J. Zaino, and J.A. Weber. 1987. Laboratory production in vivo of infectious human papillomavirus type 11. *J. Virol.* **61:** 590.

Nasseri, M., R. Hirochika, T.R. Broker, and L.T. Chow. 1987. A human papilloma virus type 11 transcript encoding an E1^E4 protein. *Virology* **159:** 433.

Saiki, R.K. S. Scharf, F. Faloona, K.B. Mullis, G.T. Horn, H.A. Erlich, and N. Arnheim. 1985. Enzymatic amplification of beta-globin genomic sequences and restriction site analysis for diagnosis of sickle cell anemia. *Science* **230:** 1350.

Polymerase Chain Reaction and Denaturing Gradient Gel Electrophoresis

R.M. Myers,[1,4] V.C. Sheffield,[2] and D.R. Cox[3,4]

Departments of [1]Physiology, [2]Pediatrics, [3]Psychiatry, and
[4]Biochemistry and Biophysics, University of California
San Francisco, California 94143

The genetic analysis of humans has been revolutionized in recent years by the ability to clone genes responsible for inherited diseases and the recognition that polymorphic DNA probes can be used as genetic linkage markers. In the majority of cases, the direct detection of mutations that cause disease and the identification of polymorphic changes in genomic sequences within and outside of genes has been accomplished by the application of the restriction-fragment-length polymorphism (RFLP) method. Although the RFLP method has been extremely fruitful, several new strategies are being developed for increasing the efficiency of detecting single-base changes in genomic DNA.

One of these approaches is denaturing gradient gel electrophoresis (DGGE), a gel system that separates DNA fragments on the basis of their melting properties (Fischer and Lerman 1983; Myers et al. 1985a,b,c; Myers and Maniatis 1986). Double-stranded DNA fragments undergo distinct melting patterns when subjected to increasing temperature or denaturant concentration; regions of the fragments (called melting domains) twenty-five to several hundred base pairs in length melt cooperatively at discrete temperatures known as T_m values. Because of the stability conferred upon DNA by stacking interactions of adjacent bases in a strand of double helix, the T_m of a melting domain is highly sensitive to differences in stacking brought about by changes in its nucleotide sequence. Alterations as small as a single-base substitution generally cause the T_m of a domain to change by a significant amount.

Fischer and Lerman (1983) designed a gel system that exploits these differences in T_m values to separate DNA fragments differing by single-base changes. The DGGE system consists of a polyacrylamide gel containing a linearly increasing gradient of DNA denaturants, such as formamide and urea, from top to bottom. As double-stranded DNA fragments enter the gel, they migrate with a mobility dependent on molecular weight. When a fragment reaches the concentration of denaturant in the gel equal to the T_m of its first melting domain, it partially melts to form a branched molecule, which slows its migration rate due to entanglement in the gel matrix. DNA fragments differing by a single-base substitution begin melting in the gel at different positions (corresponding to the T_m) and are thus separated from one another.

Two of the most valuable uses for DGGE in human genetics are in directly detecting single-base changes that cause disease and in detecting polymorphisms with DNA probes for genetic-linkage analysis. In this paper, two variations on the use of DGGE combined with the polymerase chain reaction (PCR) for these types of studies are discussed.

The PCR/GC-clamp Method

In this variation, GC-rich segments called GC-clamps are attached to genomic DNA fragments by PCR as described in Sheffield et al. (1989). As shown previously by experiment and theoretical analysis (Myers et al. 1985a,b), GC-clamps allow nearly 100% of all possible single-base changes in the attached DNA fragments to be detected by DGGE. The amplified samples are run on a denaturing gradient gel and examined by ethidium bromide staining of the gel. Besides increasing the fraction of base changes detectable in a fragment, this method overcomes the need for radioactive DNA probes because of the great increase in sensitivity resulting from PCR. To ensure that all possible base changes are being detected with this approach, it is necessary to determine the melting behavior of the DNA fragments so that optimal denaturing gradient conditions and electrophoresis times can be determined. The approaches used to determining melting behavior are described in Lerman et al. (1986), Lerman and Silverstein (1987), and Myers et al. (1987, 1988).

One of the major applications for this variation of DGGE is as a diagnostic tool for mutations or polymorphisms in relatively small genes, particularly when it is important to detect all or nearly all possible base changes in the gene. The method works best with DNA fragments in the size range of 50–500 bp. By making several sets of oligonucleotides and by PCR-amplifying in separate reactions, it is relatively straightforward to cover an entire gene of 2000–3000 bp and to detect 99% of all possible mutations in the gene. This approach is particularly suited for diagnostic tests for genes like β-globin, which is relatively small and contains many different alleles that cause disease. However, it may not be practical to use this approach to search for new polymorphisms in a probe, since such a search would be best done by quick screening of several thousand base pairs.

Some advantages of this method are the following: (1) Very high sensitivity is possible, since samples prepared from less than 10 ng of human genomic DNA are sufficient for several analyses. (2) Easy detection is possible, using ethidium bromide instead of labeled probes. (3) All possible base changes can be detected in a DNA fragment. (4) In many cases, it is possible to examine more than one test fragment in a lane, resulting in up to 1000 bp screened for base changes per lane. (5) The procedure works for test fragments containing low-, medium-, or high-copy repetitive sequences. (6) Increased resolution of different alleles is possible because of the formation of heteroduplexes. The PCR amplification results in the formation of heteroduplexes between two alleles present in a heterozygote. The heteroduplexes contain one or more single-base mismatches. This is useful because there is a large increase in separation between mismatched DNA fragments and the wild-type homoduplex. This point is discussed in Myers and Maniatis (1986) and Sheffield et al. (1989).

Some disadvantages to this method are the following: (1) It requires preliminary melting determinations for optimal resolution. (2) Oligonucleotides are expensive (one 20-mer and one 60-mer are required for each fragment to be amplified). (3) Optimal conditions for PCR vary from fragment to fragment, so some adjustments may be necessary for each set of oligonucleotides. (4) It cannot be used in multiplex fashion; each gel analyzes only one or at most two specific DNA fragments from several individuals.

The 2–3-kb PCR Method
This variation involves the amplification of a moderately large region (up to and possibly more than 3 kb) of genomic DNA by PCR, followed by digestion of the amplified fragment in separate reactions with two different frequent cutter restriction enzymes, such as *Hae*III and *Sau*3A. The digested samples are then run "blind" (i.e., with no preliminary melting determinations performed for the DNA fragments being tested) on two different denaturing gradient gels, usually with two different electrophoresis times for each sample. The gels are then exam-

179

ined by ethidium bromide staining. One gel contains a low to medium range of denaturants (e.g., 0–50% denaturant; 100% denaturant = 7 M urea + 40% formamide; see Myers et al. 1987, 1988), whereas the second gel contains a medium to high range of denaturants (e.g., 40–80% denaturant). The use of two different gels with two different electrophoresis times, coupled with the two different restriction digests, results in the screening for base changes of well over half (in many cases over 75%) of the nucleotides in the 2–3-kb DNA fragment. Because it is standardized and requires no preliminary melting behavior determinations, and because it is possible to screen several kilobases for base changes, this method is suited for searching for new polymorphisms in probes.

Advantages of this method are the following: (1) No preliminary melting behavior determinations are required; the method is standardized in that the same sets of digests, gradient conditions, and electrophoresis times are used for every test DNA fragment. (2) It screens a fairly large region of DNA for base changes with relatively little effort. (3) Very high sensitivity is possible, since samples are detected by ethidium bromide rather than by labeled probes. Note, however, it is possible to combine this approach with the use of labeled probes annealed directly to the amplified sample or in blotting experiments if desired. (4) It is easy to examine DNA fragments carrying repetitive sequences with this approach. (5) This approach also results in the formation of heteroduplexes, allowing an increase in resolution as described above.

Disadvantages of this method are the following: (1) Not all base changes are detected in the DNA fragment being tested (25–50% of the changes will be missed). (2) Oligonucleotides are expensive. (3) Optimal conditions for PCR vary from fragment to fragment, so some adjustments may be necessary for each set of oligonucleotides. (4) It also cannot be used in multiduplex fashion.

REFERENCES

Fischer, S.G. and L.S. Lerman. 1983. DNA fragments differing by single base pair substitutions are separated in denaturing gradient gels: Correspondence with melting theory. *Proc. Natl. Acad. Sci.* **80:** 1579.

Lerman, L.S. and K. Silverstein. 1987. Computational simulation of DNA melting and its application to denaturing gradient gel electrophoresis. *Methods Enzymol.* **155:** 482.

Lerman, L.S., K. Silverstein, and E. Grinfeld. 1986. Searching for gene defects by denaturing gradient gel electrophoresis. *Cold Spring Harbor Symp. Quant. Biol.* **51:** 285.

Myers, R.M. and T. Maniatis. 1986. Recent advances in the development of methods for detecting single-base substitutions associated with human genetic diseases. *Cold Spring Harbor Symp. Quant. Biol.* **51:** 275.

Myers, R.M., T. Maniatis, and L.S. Lerman. 1987. Detection and localization of single base changes by denaturing gradient gel electrophoresis. *Methods Enzymol.* **155:** 501.

Myers, R.M., V. Sheffield, and D.R. Cox. 1988. Detection of single base changes in DNA: Ribonuclease cleavage and denaturing gradient gel electrophoresis. In *Genomic analysis: A practical approach* (ed. K. Davies), p. 95. IRL Press, Oxford.

Myers, R.M., S.G. Fischer, L.S. Lerman, and T. Maniatis. 1985a. Nearly all single base substitutions in DNA fragments joined to a GC-clamp can be detected by denaturing gradient gel electrophoresis. *Nucleic Acids Res.* **13:** 3131.

Myers, R.M., S.G. Fischer, T. Maniatis, and L.S. Lerman. 1985b. Modification of the melting properties of duplex DNA by attachment of a GC-rich DNA sequence as determined by denaturing gradient gel electrophoresis. *Nucleic Acids Res.* **13:** 3111.

Myers, R.M., N. Lumelsky, L.S. Lerman, and T. Maniatis. 1985c. Detection of single base substitutions in total genomic DNA. *Nature* **313:** 495.

Sheffield, V.C., D.R. Cox, L.S. Lerman, and R.M. Myers. 1989. Attachment of a GC-clamp to genomic DNA fragments by the polymerase chain reaction results in improved detection of single base changes. *Proc. Natl. Acad. Sci.* **86:** 232.

Oligonucleotide Clamping for Complete Gene Scrutiny by Denaturing Gradient Gel Electrophoresis

L.S. Lerman and E.S. Abrams

Department of Biology, Massachusetts Institute of Technology
Cambridge, Massachusetts 02139

Although denaturing gradient gel electrophoresis is capable of detecting any single base change in a relatively long segment of DNA, perhaps up to 400 or so nucleotide pairs, it fails to respond to sequence variations in the region of the molecule with the highest T_m, that is, the most GC-dense portion. We have shown that the pattern of the thermal stability of a DNA molecule can be modified by the attachment of an extremely stable sequence ("GC clamp") to one end, such that the sequence of interest becomes the least stable part of the composite molecule, regardless of its original pattern of stability (Myers et al. 1985b). Then, nearly all base substitutions are readily demonstrable in the denaturing gradient (Myers et al. 1985a). Base substitutions can be converted into mismatches by forming heteroduplexes between the putative mutant and a wild-type molecule carrying the same clamp sequence, and all mismatches in the portion of interest will be displayed in the gradient, including single-base insertions and deletions (Lerman 1987; E. Grinfeld and L.S. Lerman, unpubl.). In previous studies, the clamp was attached to the gene sequence by recombination of the desired segment into a vector containing the clamp sequence and then excising the larger composite fragment. Cloning each genome to be tested is too laborious to be applied to a large number of genomic samples.

Practical Clamping of Gene Segments
We have examined two much simpler routes for the attachment of clamp sequences to portions of human genomic DNA. In the first (E.S. Abrams and L.S. Lerman, in prep.), a ^{32}P-labeled, single-stranded probe is prepared; this probe contains a wild-type sequence of interest adjacent to a clamp. Human genomic

DNA is digested with appropriate enzymes to produce a fragment colinear with the 3' part of the probe and then denatured and reannealed in the presence of an excess of probe. Treatment with Klenow polymerase and all four dXTPs allows extension of the 3' end of the genomic strand over the clamp region of the probe to produce a double-stranded, clamped molecule. We have shown that this technique can be used to detect virtually all the possible base changes in one part of the β-globin gene.

The second procedure for attaching a GC clamp to a genomic fragment is described by Myers et al. (this volume) and by Sheffield et al. (1989). Essentially, one of the two PCR primers has, at its 5' end, a 30-nucleotide GC-dense sequence that is not homologous to the target. The amplified molecule consists of the target sequence plus the GC clamp at one end. The paper by Myers et al. (this volume) shows that this procedure can be used to identify mutations in genomic DNA and that some of these mutations are not detectable without the clamp.

We wish to consider the extent to which either approach is appropriate for scrutiny of a complete (short) human gene. The relevant questions are the following: How many probes will be needed? What features of the sequence control the optimum choice of oligonucleotide pairs and segment boundaries? Can a number of different sequence portions be amplified simultaneously in the same reaction mixture? Can a number of different amplified sequences be examined in a single lane of a denaturing gradient gel without ambiguity as to which sequence is altered? Is the same clamp sequence appropriate for all sequence segments of the gene? The system has not yet been subjected to the exhaustive analysis that would permit definitive experimental design. Rather, we can now present a first attempt at theoretical simulation of the expected experimental results with arbitrary choices of sequences and clamps. The results of the calculation indicate that these arbitrary choices are nearly satisfactory but clearly not optimal.

Simulation of Segment Selection by PCR

We divided the human β-globin sequence into seven equal portions of 300 bp each, starting 216 bases upstream of the cap-site. Our previous studies have shown that melting is usually insensitive to sequence changes close to the ends of the molecule or immediately adjacent to the clamp; thus, each segment overlaps the previous segment by 20 bases. Previous work seems to suggest that short molecules with a clamp attached at the AT-dense end of the genomic portion may not give sharp bands in the gradient. Accordingly, clamps were attached to the higher melting end of any gene segment in the simulation. Work from both our own laboratory (E.S. Abrams and L.S. Lerman, unpubl.) and others (Sheffield et al. 1989) shows that a GC clamp of 30 bp is probably sufficient. A 45-bp clamp was used in the simulation because of our uncertainty as to the reliability of the theoretical estimate of the double-strand dissociation rate, the quantity that determines the adequacy of a clamp. The simulations were carried out using the program SQHTX as described previously (Lerman and Silverstein 1987), which provides a theoretical estimate of the difference in gradient level in terms of the equivalent temperature between a fully based-paired molecule and one containing a mismatch. The results are sampled for mismatches at every fifth position along the molecule, using an arbitrary, uniform, typical value for the degradation of helix stability attributable to the mismatch at every position. The difference in gradient level between the homoduplex and the mismatch molecule is calculated as a function of running time in the gel; the program also reports the anticipated depth of the homoduplex fragment.

The results for the seven segments are shown in Figure 1, where panel A gives the expected displacements for a relatively short period of electrophoresis in a 60°C bath, and panel B shows the displacements after a longer time in a 70°C bath. It is convenient to report the gradient shifts in terms of equivalent temperature, since the actual geometrical distance in the gel can be adjusted according to the gradient parameters. A shift of 0.1°C indicated in the figure would correspond to about 1.5 mm in a gel, which is easily resolved. It is seen that shifts much larger than 0.1°C result from mismatches almost everywhere in the β-globin gene, if both sets of conditions are included, and that most mismatches are expected to give displacements of 0.5 cm or more. Figure 2 shows the band positions to be expected in a denaturing gradient for normal gene segments (heavy bands) and mutants with mismatches at peak sites (lighter bands) under conditions corresponding to panel A of Figure 1. Only one pair of the seven segments might not be distinguished. As demonstrated in our laboratory and elsewhere, multiple fragments can be included in the same lane. However, at least two reaction mixtures and two lanes of a gel would be needed for the comprehensive scrutiny of any gene by

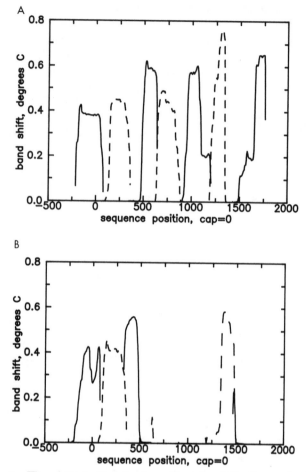

Figure 1 The shift in the final gradient level for heteroduplex between mutant and normal human β-globin strands. The ordinate shows the difference in retardation level from that of normal homoduplexes as a function of the site of variation in each of seven segments carrying a GC-dense clamp sequence at one end. Position 0 is the base immediately upstream of the cap site. The calculations follow the theory given in Lerman et al. (1984). The fragments have 45-bp clamps attached at either the 5′ or the 3′ end as follows, left to right: 3′, 5′, 5′, 5′, 3′, 3′, and 5′. (*A*) Electrophoresis for about 8 hours in an environment of 60°C. (*B*) The same molecules at about 20 hours in an environment of 70°C. The lines are alternately solid or dashed for clarity.

PCR to prevent synthesis between inappropriate pairs of primer oligonucleotides. The lines on the right in Figure 2 show simulated mixtures of the odd- and even-numbered segments.

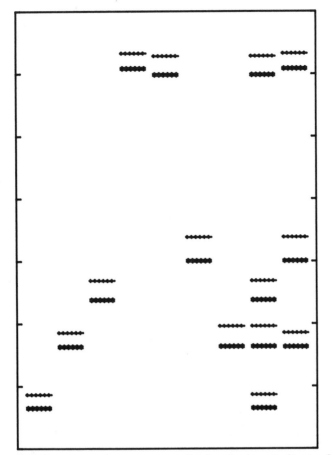

Figure 2 The expected appearance of a denaturing gradient gel carrying bands of each of the seven segments indicated in Figure 1A. The two lanes at the right show sample mixtures of the odd- and even-numbered segments. The gradient simulates a gel with a urea-formamide gradient running from an equivalent temperature of 62°C at the top to 76°C at the bottom. The heavy bands indicate the position of the normal homoduplex, and the lighter bands indicate the position of a heteroduplex carrying a typical mismatch at the peak sequence position. Actual displacements depend on the details of the mismatch and may be greater or lesser than those indicated here.

Normal and mutant levels remain clearly distinguishable within each mixture.

The simulation reveals the need for more deliberate experimental design. The choice of sequence boundaries has resulted in the exclusion of a short region near the center of the second intron (near 800) from sensitivity to mismatches. The need for two different electrophoresis conditions implies additional labor, limiting the number of genomic samples that can be examined. Although it would be desirable to examine segments longer than 300 bp, it appears that mismatches in some portions of the gene may fall below the minimum resolvable limit with longer segments.

The simulation is equally applicable to fragments produced by PCR or by the clamp extension method. The sensitivity of PCR obviates the need for blood samples and frees the test from dependence on favorably positioned restriction sites. The clamp extension method is more rapid; it does not require a priori knowledge of the gene sequence; it is unaffected by the fidelity of the polymerase in amplification; and it can reveal methylation, even of a single base, or other base modifications as well as sequence variation.

ACKNOWLEDGMENTS

We are grateful to Susan Murdaugh for her technical assistance. This work was supported by a grant from the National Institutes of Health.

REFERENCES

Lerman, L.S. 1987. Detecting sequences in a gene. *Somatic Cell Mol. Genet.* **13**: 419.

Lerman, L.S. and K. Silverstein. 1987. Computational simulation of DNA melting and its application to denaturing gradient gel electrophoresis. *Methods Enzymol.* **155**: 482.

Lerman, L.S., S.G. Fischer, I. Hurley, K. Silverstein, and N. Lumelsky. 1984. Sequence-determined DNA separations. *Annu. Rev. Biophys. Bioeng.* **13**: 399.

Myers, R.M., S.G. Fischer, L.S. Lerman, and T. Maniatis. 1985a. Nearly all single base substitutions in DNA fragments joined to a GC-clamp can be detected by denaturing gradient gel electrophoresis. *Nucleic Acids Res.* **13**: 3131.

Myers, R.M., S.G. Fischer, T. Maniatis, and L.S. Lerman. 1985b. Modification of the melting properties of duplex DNA by attachment of a GC-rich DNA sequence as determined by denaturing gradient gel electrophoresis. *Nucleic Acids Res.* **13**: 3111.

Sheffield, V.C., D.R. Cox, L.S. Lerman, and R.M. Myers. 1989. Attachment of a 40 base pair G + C-rich sequence (GC-clamp) to genomic DNA fragments by the polymerase chain reaction results in improved detection of single base changes. *Proc. Natl. Acad. Sci.* **86**: 232.

Detection of Single-base Mismatches in PCR Products Using the Solution Melting Method

F.I. Smith and T. Latham

Department of Microbiology, Mount Sinai School of Medicine
New York, New York 10029

Direct sequencing of polymerase-chain-reaction (PCR)-amplified products unequivocally identifies previously unknown point mutations. However, it is often useful to have a method for rapidly screening multiple bases for the approximate location of such mutations before sequencing. RNase-A cleavage of RNA-DNA heteroduplexes has been successfully used to detect mutations in PCR products (see, for example, Almoguera et al. 1988), but it cannot detect all mutations because its efficiency varies depending on the mismatch and its context (Myers et al. 1985). Denaturing gradient gel electrophoresis of double-stranded (ds) DNA heteroduplexes is capable of detecting all single-base mismatches that occur in the low-melting domains (LMD) of ds DNA molecules, but it loses its resolution when strand dissociation occurs (Lerman et al. 1984). Myers and co-workers (Sheffield et al. 1989; Myers et al., this volume) have recently described one approach to overcome this limitation: PCRs are performed using one primer that contains an additional GC-rich sequence at its 5' end, which functions as a clamp at one end of the resultant PCR product. The solution melting method (Smith et al. 1988; Latham and Smith 1989) provides a different approach to detecting mutations in the high-melting domains (HMD) of PCR products that is based on melting heteroduplexes in solution, followed by polyacrylamide gel electrophoresis to monitor for strand dissociation.

Melting of ds nucleic acids proceeds step-wise through a series of discrete domains as the denaturant concentration increases. The conditions under which strand dissociation occurs are determined by the sequence of the HMD, irrespective of the length or composition of the other domains that may be present. Therefore, when a mixture of radioactively labeled ds RNA-DNA duplexes of different length that contain identical HMDs but different LMDs are heated in a series of tubes containing varying concentrations of formamide, they undergo strand dissociation at the same concentration of formamide. Strand dissociation can be detected by polyacrylamide gel electrophoresis and subsequent autoradiography, which identifies the radiolabeled probe in either a ds (fast migrating) or single-stranded (ss [slow-migrating]) form. This method is sensitive enough to detect destabilization of a HMD in an RNA-DNA heteroduplex by a single-base mismatch, as revealed by the earlier disappearance of the relevant ds species in autoradiograms (Latham and Smith 1989).

Detection of a Polymorphism in the Factor VIII Gene

We have investigated whether the solution melting method could also detect single-base mismatches in PCR products. During the process of PCR amplification, the polymerase will introduce errors into the newly synthesized DNA, and any such errors will be subsequently amplified, resulting in the accumulation of mutated sequences. After 30 cycles of amplification using *Thermus aquaticus* polymerase, approximately 1 introduced mutation per 400 nucleotides is found in the amplified fragments (Saiki et al. 1988). Therefore, the analysis of PCR fragments containing HMDs approximately 130 bp and shorter (which has been shown to be the upper limit of HMD length in RNA-DNA heteroduplexes that can be reliably analyzed by this method; Latham and Smith 1989) should be unaffected by the introduction of mutations during amplification.

To test this system, the detection of a previously characterized *Bcl*I-cleavage-site polymorphism in the factor VIII gene (Gitschier et al. 1985) was attempted. The experimental protocol is shown diagrammatically in Figure 1. PCRs were performed on genomic DNA samples obtained from males that have different alleles (a1 or a2) at this polymorphic site, thus generating DNA duplexes differing from each other by a single point mutation. The PCR product obtained from one allele was digested with the restriction enzyme *Dra*I to truncate its LMD, thereby producing DNA duplexes of different length. Both sets of PCR products were then hybridized with a radiolabeled RNA probe complementary to only one of these alleles, generating ei-

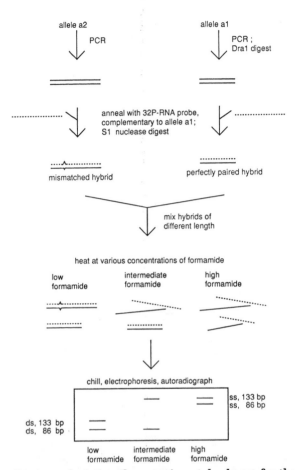

Figure 1 Diagram depicting the experimental scheme for the analysis of mutations in PCR-amplified DNA by the solution melting method. The factor VIII gene from base positions 13,679 to 13,812 was PCR amplified, using genomic DNA from males known to contain different *BclI* polymorphisms (allele a1 or allele a2). Half of one PCR product was then digested with *DraI* (in the example shown here, allele a1). Both digested and undigested products were then hybridized to radiolabeled RNA probes complementary to one allele. Hybrids were then digested with S1 nuclease to remove unreacted probe and to generate blunt-ended duplexes. Different combinations of duplexes were then mixed together, aliquoted into a series of tubes containing varying formamide concentrations, heated briefly, and rapidly chilled. Samples are then fractionated according to size and form, by polyacrylamide gel electrophoresis, and detected by autoradiography. Duplexes containing mismatches in their HMD will undergo strand dissociation at lower concentrations of formamide than will perfectly matched duplexes. (Reprinted, with permission, from Latham and Smith 1989.)

191

ther a perfectly matched duplex or a duplex containing a single-base mismatch in its HMD. Solution melting analysis of varying mixtures of these different length hybrids shows that those derived from DNA carrying the same allele melt indistinguishably (Fig. 2A), whereas those that carry different alleles melt at different concentrations of denaturant (Fig. 2B,C). Therefore, the presence of a single-base mutation in an X-linked gene was easily detectable by solution melting analysis of PCR-amplified genomic DNA obtained from male individuals. In contrast, the detection of mutations in heterozygous autosomal genes, where the loss of half the intensity of a band on a gel must be detected, is more difficult. Subjective analysis by eye was found to be unreliable, but the presence of muta-

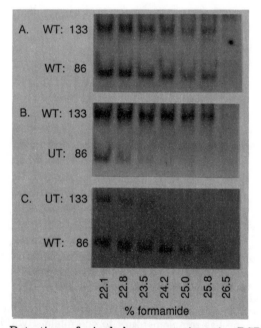

Figure 2 Detection of single-base mutations in PCR-amplified genomic DNA. Method as outlined in Fig. 1. (*A*) Hybrids derived from genomic DNA bearing the same polymorphism melt indistinguishably. Both 133-bp and 86-bp RNA-DNA hybrids are derived from individuals bearing allele a1. (*B* and *C*) Hybrids derived from genomic DNA bearing different polymorphisms melt at different concentrations of formamide. (*B*) The 133-bp hybrid is derived from individuals bearing allele a1, and the 86-bp hybrid is derived from individuals bearing allele a2. (*C*) The situation is the reverse of *B*. (Reprinted, with permission, from Latham and Smith 1989.)

tions could be detected by scanning autoradiographs by densitometry. However, because this is tedious, this method may be more useful for the analysis of X-linked genes.

Potential Application for Genetic Disease Analysis

Because of the length limitations of this method, it is not suitable for screening large regions of the human genome for polymorphisms. However, it seems well suited to the simultaneous screening of multiple exons for disease-causing mutations. Exons show in general about a 10% higher GC concentration than do introns (Bernardi et al. 1985), and thus coding sequences, particularly those in short exons, may generally lie in HMDs. Additionally, most genes contain relatively short exons. For example, the factor VIII gene contains 26 exons: 4 exons shorter than 94 bp, 12 exons 117–177-bp long, 7 exons 213–262-bp long, 1 exon of 313 bp, and 2 exons longer than 1900 bp (Gitschier et al. 1984). Thus, 24 of the 26 exons are quite short and are very likely to contain HMDs of 130 bp or less. Therefore, PCR amplification of selected groups of these exons (selected such that the resulting hybrids differ sufficiently in length as to be distinguishable on gel electrophoresis), followed by hybridization with a radioactively labeled RNA probe and subsequent simultaneous analysis by the solution melting method, should identify which exons contain mutations in their HMDs. Subsequent screening by denaturing gradient gel analysis should detect any other mutations that occur in exons containing a LMD. Alternately, to increase the likelihood of detecting all mutations in the HMDs, without relying on the probability that most HMDs will be 130 bp or less, ds RNA hybrids could be analyzed. PCRs could be primed using oligonucleotides containing a T7 RNA polymerase promoter, followed by transcription with T7 polymerase to produce ss RNA, and subsequent hybridization with an RNA probe. Since 22 of 26 exons are under 230-bp long and the solution melting method can reliably detect single-base mismatches in ds RNA duplexes containing HMDs of up to 250-bp long (Latham and Smith 1989), a combination of the solution melting analysis and denaturing gradient gel electrophoresis should detect 100% of single-base mutations in these 22 exons, as well as having a high probability of detecting all mutations in the remaining 2 short exons (262-bp and 313-bp long). Most genes so far characterized also contain predominantly short exons and usually fewer exons than are present in the factor VIII gene. In conclusion, the solution melting method is a valuable addition to the repertoire of techniques capable of detecting single-base mutations in DNA molecules, and it should prove most useful in the characterization of disease-causing mutations in X-linked genes.

ACKNOWLEDGMENTS
We thank Dr. Jane Gitschier for advice on PCR amplification of the factor VIII gene. The project was aided by National Institutes of Health grant DK-38381 and by Basil O'Connor Starter Scholar Research Award no. 5-646 from the March of Dimes Birth Defects Foundation.

REFERENCES
Almoguera, C., D. Shibata, K. Forrester, J. Martin, N. Arnheim, and M. Perucho. 1988. Most human carcinomas of the exocrine pancreas contain mutant c-K-*ras* genes. *Cell* 53: 549.

Bernardi, G., B. Olofsson, J. Filipski, M. Zerial, J. Salinas, G. Cuny, M. Meunier-Rotival, and F. Rodier. 1985. The mosaic genome of warm-blooded vertebrates. *Science* 228: 953.

Gitschier, J., D. Drayna, E.G.D. Tuddenham, R.L. White, and R.M. Lawn. 1985. Genetic mapping and diagnosis of hemophilia A achieved through a Bcl1 polymorphism in the factor VIII gene. *Nature* 314: 738.

Gitschier, J., W.I. Wood, T.M. Goralka, K.L. Wion, E.Y. Chen, D.H. Eaton, G.A. Vehar, D.J. Capon, and R.M. Lawn. 1984. Characterization of the human factor VIII gene. *Nature* 312: 326.

Latham, T. and F.I. Smith. 1989. Detection of single base mutations in DNA molecules using the solution melting method. *DNA* 8: 223.

Lerman, L.S., S.G. Fischer, I. Hurley, K. Silverstein, and N. Lumelsky. 1984. Sequence-determined DNA separations. *Annu. Rev. Biophys. Bioeng.* 13: 399.

Myers, R.M., Z. Larin, and T. Maniatis. 1985. Detection of single base substitutions by ribonuclease cleavage at mismatches in RNA:DNA duplexes. *Science* 230: 1242.

Saiki, R.K., D.H. Gelfand, S. Stoffel, S.J. Scharf, R. Higuchi, G.T. Horn, K.B. Mullis, and H.A. Erlich. 1988. Primer-directed enzymatic amplification of DNA with a thermostable DNA polymerase. *Science* 239: 487.

Sheffield, V., D.R. Cox, L.S. Lerman, and R.M. Meyers. 1989. Attachment of a 40-base-pair G + C-rich sequence (GC-clamp) to genomic DNA fragments by the polymerase chain reaction results in improved detection of single-base changes. *Proc. Natl. Acad. Sci.* 86: 232.

Smith, F.I., T.E. Latham, J.A. Ferrier, and P. Palese. 1988. Novel method of detecting single base substitutions in RNA molecules by differential melting behaviour in solution. *Genomics* 3: 217.

Genetic Mapping Using the Polymerase Chain Reaction on Single Sperm

H. Li,[1] M. Boehnke,[2] F.S. Collins,[3] and N. Arnheim[1]

[1]Department of Biological Sciences, University of Southern California
Los Angeles, California 90089-0371

Departments of [2]Biostatistics, [3]Internal Medicine, Human Genetics,
and the Howard Hughes Medical Institute, University of Michigan
Ann Arbor, Michigan 48109

The construction of genetic maps in higher organisms depends on the ability to analyze the progeny of selected matings or to compute linkage relationships by means of pedigree analysis. In humans, only the latter is possible. Using restriction-fragment-length polymorphisms (RFLPs), there has been significant progress toward the construction of a human linkage map (for reviews, see Donis-Keller et al. 1987; White et al. 1987; White and Lalouel 1987). Pedigree analysis is thought to be able to measure genetic distances to a resolution of approximately 1 cM (encompassing ~1000 kb of DNA) with statistical reliability (Aston et al. 1988). The analysis of smaller genetic distances requires an examination of such a large number of individuals from informative families that it is impractical. Recently a new approach to measure recombination frequencies between human nucleotide sequence polymorphisms, using a method that does not rely on family studies, was proposed (Li et al. 1988). This approach is based on directly determining whether individual sperm are of the parental or recombinant genotype using the in vitro gene amplification procedure known as the polymerase chain reaction (Saiki et al. 1985, 1988; Mullis and Faloona 1987). Dividing the number of recombinant sperm determined by this procedure by the total number of sperm examined will provide an estimate of the frequency of recombination. Studying thousands of sperm in this way would allow us to generate fine-structure genetic maps at a resolution far greater than that possible using pedigree analysis.

Our ability to type individual sperm at the DNA level will provide a fundamentally new and alternative approach to determining the physical order of DNA polymorphisms that are so tightly linked that they cannot be resolved by pedigree analysis. The linear order of genetic markers on a chromosome can be inferred by carrying out three point crosses according to classic methods used in experimental genetics (see Sturtevant and Beadle 1939). Ordering genes using sperm typing might in some cases be even faster than the physical mapping procedures. This may be especially significant in the case of random polymorphisms tightly linked to disease-causing loci. Because our method will allow us to examine very large numbers of meiotic products, even tightly linked polymorphisms could be ordered with respect to one another by three point crosses. Such fine-structure maps might prove of great value in attempts to locate the disease-causing locus itself. A mathematical strategy for determining gene order in three point crosses using data on single sperm will be presented.

We will also summarize our existing data and discuss our progress in solving the so-called false recombinant problem, which affects the level of resolution of the single-sperm mapping procedure. False recombinants can result from the simultaneous occurrence of two experimental errors: a sperm sample containing not one sperm but two and inefficient target amplification. Calculations to estimate the frequency of false recombinants will also be considered.

The single-sperm typing protocol will, in addition, be valuable in studies on recombination hot spots and in measuring the relationship between physical distance and recombination frequency in man. The application of this technology to single cells other than sperm will make it possible to study cell-to-cell variation in developmental processes that involve DNA rearrangements or other genetic alterations. Prenatal diagnosis carried out on a single cell derived from a preimplantation embryo resulting from in vitro fertilization is also possible.

REFERENCES

Aston, C.E., S.L. Sherman, N.E. Morton, P.W. Speiser, and M.I. New. 1988. Genetic mapping of the 21-hydroxylase locus: Estimation of small recombination frequencies. *Am. J. Hum. Genet.* **43**: 304.

Donis-Keller, H., P. Green, C. Helms, S. Cartinhour, B. Weiffenbach, K. Stephens, T.P. Keith, D.W. Bowden, D.R. Smith, E.S. Lander, D. Botstein, G. Akots, K.S. Rediker, T. Gravius, V.A. Brown, M.B. Rising, C. Parker, J.A. Powers, D.E. Watt, E.R. Kauffman, A. Bricker, P. Phipps, H. Muller-Kahle, T.R. Fulton, S. Ng, J.W. Schumm, J.C. Braman, R.G. Knowlton, D.F. Barker, S.M. Crooks,

S.E. Lincoln, M.J. Daly, and J. Abrahamson. 1987. A genetic linkage map of the human genome. *Cell* **51**: 319.

Li, H., U.B. Gyllensten, X. Cui, R.K. Saiki, H.A. Erlich, and N. Arnheim. 1988. Amplification and analysis of DNA sequences in single human sperm and diploid cells. *Nature* **335**: 414.

Mullis, K.B. and F.A. Faloona. 1987. Specific synthesis of DNA in vitro via a polymerase catalysed chain reaction. *Methods Enzymol.* **155**: 335.

Saiki, R.K., S. Scharf, F. Faloona, K.B. Mullis, G.T. Horn, H.A. Erlich, and N. Arnheim. 1985. Enzymatic amplification of beta-globin genomic sequences and restriction site analysis for diagnosis of sickle cell anemia. *Science* **230**: 1350.

Saiki, R.K., D.H. Gelfand, S. Stoffel, S.J. Scharf, R. Higuchi, G.T. Horn, K.B. Mullis, and H.A. Erlich. 1988. Primer directed enzymatic amplification of DNA with a thermostable DNA polymerase. *Science* **239**: 487.

Sturtevant, A.H. and G.W. Beadle. 1939. Chromosome maps. In *An introduction to genetics*, p. 93. Dover Publications, New York.

White, R. and J.-M. Lalouel. 1987. Investigation of genetic linkage in human families. *Adv. Hum. Genet.* **16**: 121.

White, R., J.-M. Lalouel, P. O'Connell, Y. Nakamura, M. Leppert, and M. Lathrop. 1987. Current status in mapping the human genome: 440 RFLPs in 59 families and 690 new RFLPs. *Cytogenetic Cell Genet.* **46**: 715.

Detection of Targeted Gene Modifications by Polymerase Chain Reaction

O. Smithies, B.W. Popovich, and H.-S. Kim

Department of Pathology, University of North Carolina
at Chapel Hill, North Carolina 27599-7525

The ultimate goal of our experiments is to be able to modify any type of mammalian gene in a predetermined way. In attempting to do this, we have been using gene targeting. Targeting depends on introducing DNA into a cell so that the exogenous DNA matches sequences in the target gene. This matching allows homologous recombination to occur between the incoming and the target sequences. Depending on the arrangement of the incoming sequences, the recombination can lead either to their insertion into the target locus or to replacement of sequences in the target by the incoming DNA (Gregg and Smithies 1986). We have obtained both types of event.

Methods to Detect Gene Targeting

Detection and hence isolation of cells containing a correctly modified target locus is possible in three ways that have different levels of usefulness. At the first level, the target locus must have a directly selectable phenotype. We have succeeded at this level with the hypoxanthine-guanine phosphoribosyltransferase (*HPRT*) locus (Doetschman et al. 1987, 1988). Clearly, there are only a limited number of loci that can be arranged to provide a directly selectable phenotype.

At the second level, the target locus must still be active in the treated cells, but artificial selection, for example, via a drug resistance marker can be used to allow at least enrichment of the treated cells for transformants. We used this method in our early experiments with the human β-globin locus (Smithies et al. 1985), but identification of correctly modified cells among the many transformants still required a laborious second screening process. Recently, Mansour et al. (1988) have devised an interesting variation of this principle that greatly reduces the labor of the second screening process used to find correctly modified cells. The new method however still requires the insertion of a drug-resistance gene into the target locus, which in turn must permit adequate expression of the inserted drug-resistance gene.

At the third level are target genes for which the planned modification either cannot be made artificially selectable (perhaps because the locus will not support expression in the target cells) or is of a type where the addition of a drug-resistance gene is undesirable. It is for the detection of targeted genes of this third category of modification that we have been using the polymerase chain reaction (PCR).

PCR Detection Method

The principle of the detection method, which is essentially a screening assay, depends on arranging the target and incoming DNA sequences in such a way that a particular PCR-amplifiable fragment is only found in cells in which the correct targeted modification has occurred. If AHB represents the target locus, and CHD represents the incoming DNA (with H being sequences common to them that allow the homologous recombination), then the recombinant will be AHD or CHB. By choosing a primer A, specific for the target locus, and a primer D, specific for the incoming DNA, fragment AHD will only be amplified when the targeting has been successful.

Kim and Smithies (1988) tested this assay procedure by first working out PCR conditions that allow the detection of single cells containing a single recombinant gene. The recombinant fragment AHD in their example was 400-bp long, and after amplification by PCR it could be seen as a visible band in an ethidium-bromide-stained agarose electrophoresis gel. The recombinant fragment could still be detected when a few correctly modified cells were mixed with a 10,000-fold excess of unmodified target cells. Suitable control experiments showed that the PCR method itself did not generate artefactual recombinants under conditions more stringent than the assay was likely to encounter.

Application of the PCR Assay

The first system for which we are using this PCR assay for detecting targeted modifications is illustrated in Figure 1. The experiment is designed to correct a human sickle cell β-globin gene on human chromosome 11 by replacing the defective gene with an incoming fragment of DNA from a normal β-globin sequence. Clearly, in this case, it is completely inappropriate to insert any selectable marker into the locus, and so the PCR

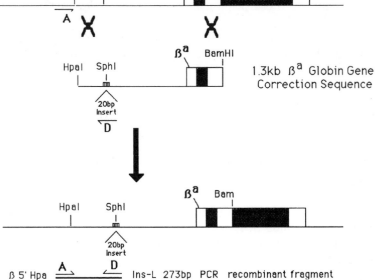

β^S Globin Gene

Figure 1 The top line represents the endogenous β^S-globin gene target with the incoming microinjected 1.3-kb β^A-globin correction sequence below. The Xs between these two sequences represent potential cross-over points in homologous DNA regions. The product of gene targeting is shown below the vertical arrow. The correctly targeted gene can be detected by PCR, with the 273-bp recombinant fragment, shown at the bottom, becoming diagnostic for the desired recombination event. The locations of the sequences used for the primers (A and D), specific for the target locus and the incoming DNA, respectively, are shown by the two horizontal arrows.

method of detecting recombinants is being used. However at present (although probably not in the future), we have had to introduce 20 nucleotides into the incoming DNA to serve as primer D. (Primer A is chosen from sequences in the target locus that are not matched in the incoming DNA.) The total length of homology in the illustrated example is about 1.3 kb, although only 200 bp correspond to H in the amplified 273-bp recombinant fragment AHD.

Currently, our procedure is to microinject a few molecules of the correcting sequence into the nuclei of most of the cells in a pool of about 200 target cells. The target cells are mouse erythroleukemia cells containing a single copy of human chromosome 11 carrying the sickle cell β-globin gene mutation. A number of pools are injected, and the cells are then allowed to grow. Cells are taken from each pool and are subjected to amplification to identify those pools in which the recombinant fragment AHD is present.

Candidate pools containing the recombinant fragment have been identified, and we are now proceeding to subclone cells from these positive pools to establish that they include bona fide recombinants and to obtain pure colonies of the correctly modified cells. We also need to be sure that we are not being misled by any of the types of contamination that all of us who work with PCR at this level of sensitivity have at some time encountered.

The Future

We hope to improve this approach to detecting gene targeting. In particular, we hope and expect that we will be able to detect targeted gene modifications in which not even 20 nucleotides need be added, but rather modified cells will be detected by the modification itself, even when it is only a single nucleotide.

ACKNOWLEDGMENTS

This investigation was supported by a National Institutes of Health National Research Service award (6M-12285) and National Cooley's Anemic Foundation award (B.P.) and National Institutes of Health grants GM-20069 and HL-37001 (O.S.).

REFERENCES

Doetschman, T., N. Maeda, and O. Smithies. 1988. Targetted mutation of the *HPRT* gene in mouse embryonic stem cells. *Proc. Natl. Acad. Sci.* **85**: 8583.

Doetschman, T., R.G. Gregg, N. Maeda, M.L. Hooper, D.W. Melton, S. Thompson, and O. Smithies. 1987. Targetted correction of a mutant *HPRT* gene in mouse embryonic stem cells. *Nature* **330**: 576.

Gregg, R.G. and O. Smithies. 1986. Targeted modification of human chromosomal genes. *Cold Spring Harbor Symp. Quant. Biol.* **51**: 1093.

Kim, H.-S. and O. Smithies. 1988. Recombinant fragment assay for gene targetting based on the polymerase chain reaction. *Nucleic Acids Res.* **16**: 8887.

Mansour, S.L., K.R. Thomas, and M.R. Capecchi. 1988. Disruption of the proto-oncogene *int-2* in mouse embryo-derived stem cells: A general strategy for targetting mutations to non-selectable genes. *Nature* **336**: 348.

Smithies, O., R.G. Gregg, S.S. Boggs, M.A. Koralewski, and R.S. Kucherlapati. 1985. Insertion of DNA sequences into the human chromosomal β globin locus via homologous recombination. *Nature* **317:** 230.

Long-range Ordering of Clones by Probe Fingerprinting

A. Craig, D. Nizetic, J.D. Hoheisel, A.P. Monaco, G. Zehetner, G. Lennon, B. Young, and H. Lehrach

Imperial Cancer Research Fund Laboratories, Lincoln's Inn Fields
London WC2A 3PX, United Kingdom

The mammalian genome is expected to contain on the order of 100,000 genes responsible for the formation, shape, and, at least to some extent, behavior of mammals. Although a few of these genes have been identified, mostly using information from a previously characterized gene product, the vast majority of the genes of the mammalian genome remain to be identified. Two main routes are expected to provide access to these genes, for most of which no gene product is known or can realistically be expected to be identified. One approach, already used successfully to identify genes responsible for both human genetic diseases and mouse developmental mutations, uses genetic information derived from the analysis of mutants in the genes to clone and identify the mutated gene sequences. Again, this approach is unlikely to be completely general, since many mutations will not be identifiable by either human or mouse genetics, limited respectively by the requirements for large families and easily scorable phenotypes. More generally applicable might be the molecular analysis of the entire genome, allowing the identification of coding regions, e.g., by the analysis of sequences conserved between different species, reflecting a selection against random mutations and therefore a probable function of the identified sequence.

To simplify both the identification of genes defined by mammalian mutations and the direct molecular analysis of large regions of mammalian genomes, we have been developing a potentially quite rapid approach for the construction of ordered clone libraries, based on the detection of overlapping clones by hybridization of the clones to a large number (50–200) of probes selected to hybridize to 10–50% of all clones. Different types of probes fulfilling this condition could be used (short oligonucleotides, cloned middle repetitive sequences, pools of cDNA clones,

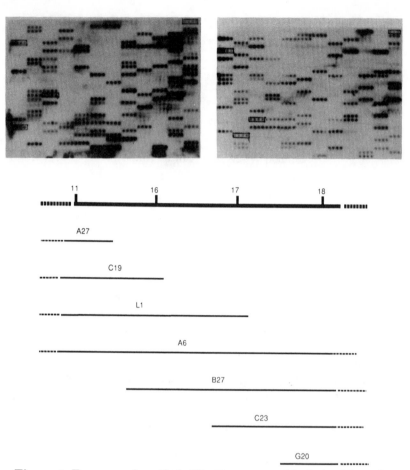

Figure 1 Two examples of hybridizations to cosmids derived from the HSV genome. (*Top*) Each cosmid is defined by four dots in an array with 32 rows and 12 columns, each column consisting of four dots. (*Bottom*) The partial cosmid map covered by the oligos used in the hybridizations.

or clone end fragments), although the main emphasis of our work up to now has been directed toward the use of oligonucleotides or pools of oligonucleotides as hybridization probes.

Figure 2 A diagram showing the arrangement of the four isomers of the HSV genome as detected by the oligofingerprinting. Overlapping cosmids covering the entire HSV genome are shown in the lower half of the diagram. Cosmids E10, C19, A6, and F3 are repeated for each isomer as they cannot be assigned uniquely.

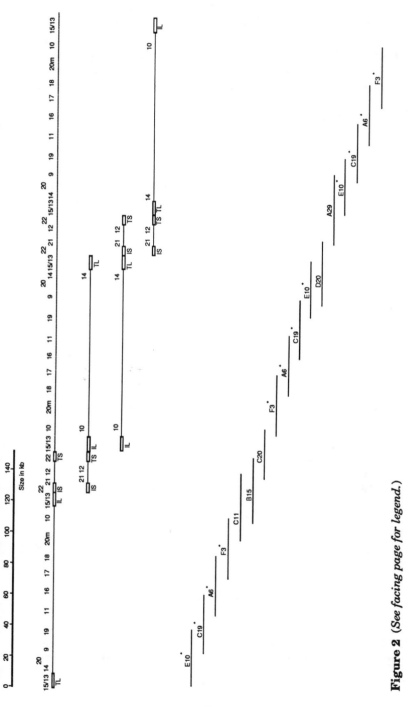

Figure 2 (See facing page for legend.)

207

Overlapping clones can be expected to show very similar patterns of hybridization. Similar to the situation in genetic linkage mapping, these types of binary (or grey level) data can at least in principle be used to identify overlapping clones and to establish the linear order of both clones and probes on the genome (Michiels et al. 1987).

RESULTS

After testing the basic feasibility of the approach by computer simulations and demonstrating specific hybridization of, e.g., dodecanucleotides to DNA from phage plaques, spotted phage supernatants, and most recently cosmid clones, work has been carried out on the development and testing of many aspects of this approach. To test the ordering of clones by hybridization in an easily verifiable situation, we constructed a cosmid library from the (completely sequenced) genome of herpes simplex virus type-1 (HSV-1; 153 kb; McGeoch et al. 1988), picked 350 clones (~100-fold coverage) into microtiter plates, and hybridized robot-spotted high-density replicas of the library with a series of oligonucleotide probes (dodecanucleotides) covering the entire genome. Figure 1 shows a set of hybridizations of probes to a set of clones and demonstrates the expected changes in hybridization patterns between cosmids located in adjacent positions on the genome.

The interpretation of the hybridization results, an order of cosmids on a linear arrangement of the four isomeric forms of HSV-1 (Delius and Clements 1976), is shown in Figure 2. Work is now under way to use this system to develop and test the procedures used in the automated analysis of these data (image analysis of the films to identify and score colonies and programs carrying out the reconstruction of the clone order) on this data set.

In addition to the tests of the procedure mentioned above, cosmid (~15 chromosome equivalents) and large insert λ-libraries (~100 chromosome equivalents) have been constructed from flow-sorted human X chromosomes. Fractions of the libraries have been picked into microtiter plates and spotted in multiple replicas as similar high-density arrays of clones (~10,000 clones/22 x 22-cm filter). In addition to their intended use in oligofingerprinting, these high-density chromosome-specific libraries have been screened with a number of X-chromosomal probes and have proven highly useful in chromosome walking experiments, since hybridization of two filters

(20,000 clones, 4–5 chromosome equivalents) allows immediate identification of the corresponding cosmid clones stored in microtiter plates. Similarly, construction of a library of X-chromosomal sequences in yeast artificial chromosome (YAC) vectors from a cell line containing a human X chromosome in a hamster background has been started. Clones covering approximately one fifth of the chromosome have been identified and picked into microtiter plates.

We expect these and similar libraries, used in the identification of cosmid clones and YAC clones from genetically mapped probes and in the intended ordering of cosmid clones (and possibly also YAC clones) by the use of the hybridization protocol described above, to serve as references, ultimately connecting genetic, physical, and transcriptional maps of individual chromosomes.

SUMMARY AND DISCUSSIONS

A series of experimental problems arising in establishing the long-range order of clones by oligonucleotide fingerprinting have been addressed with encouraging results. Cosmid, λ-, and YAC libraries of the X chromosome have been constructed in preparation for analysis by such an ordering protocol and have already proven their usefulness in the rapid isolation of clones from the X chromosome. The use of such high-density replicas of primary libraries as a reference in linking genetic, physical, and transcriptional maps of specific chromosomes is proposed (and planned).

ACKNOWLEDGMENTS

We thank Peter Goodfellow for discussions on the feasibility of the reference library concept, Nathan Ellis and Annemarie Poustka for providing information on cosmid clones identified in the X-chromosome library, Annemarie Frischauf for comments on the manuscript, and Gill Bates and Zoia Larin for discussions and help in cloning in YAC vectors. A long-term European Molecular Biology Organization fellowship to J.D.H. is gratefully acknowledged.

REFERENCES

Delius, H. and J.B. Clements. 1976. A partial denaturation map of herpes simplex virus type 1 DNA: Evidence for inversions of the unique DNA regions. *J. Gen. Virol.* **33**: 125.

McGeoch, D.J., M.A. Dalrymple, A.J. Davison, A. Dolan, M.C. Frame, D. McNab, L.J. Perry, J.E. Scott, and P. Taylor. 1988. The complete DNA sequence of the long unique region in the genome of herpes simplex virus type 1. *J. Gen. Virol.* **69:** 1531.

Michiels, F., A.G. Craig, G. Zehetner, G.P. Smith, and H. Lehrach. 1987. Molecular approaches to genome analysis: A strategy for the construction of ordered overlapping clone libraries. *Comput. Appl. Biosci.* **3:** 203.

Interfacing In Vitro DNA Amplification with DNA Sequencing Strategies

L.J. McBride,[1] S.M. Koepf,[1] R.A. Gibbs,[2]
P. Nguyen,[2] W. Salser,[3] P.E. Mayrand,[1]
M.W. Hunkapiller,[1] and M.N. Kronick[1]

[1]Applied Biosystems, Foster City, California 94404
[2]Baylor College of Medicine, Houston, Texas 77030
[3]Department of Biology, University of California
Los Angeles, California 90024

The advent of the dideoxy termination method of DNA sequencing (Sanger et al. 1977) has encouraged scientists to devise rational strategies to tackle large sequencing projects (Hood et al. 1987). Clearly, development of high-throughput methods suitable for automation is a rational focus for the future. Toward this goal, automated DNA sequence analysis has been achieved by combining fluorescence-based detection with dideoxy termination (Smith et al. 1986). However, other aspects of a sequencing project such as cloning, template preparation, and sequencing reactions are complex manual operations that require significant time and skill. Furthermore, the types of manipulations required for these manual methods vary with the type of template being sequenced and therefore lack the repetitive nature conducive for automation.

Polymerase chain reaction (PCR), the in vitro DNA amplification scheme (Mullis and Faloona 1987), may offer a coherent strategy for automating "front end" sequencing steps (Stoflet et al. 1988; Innis et al. 1988). We also believe PCR amplification has tremendous potential to simplify many DNA sequencing and template preparation methods, and now we present our progress toward interfacing PCR technology with the fluorescence-based dideoxy termination method of DNA sequencing.

A Simple PCR-DNA Sequencing Method

A limitation of many Sanger DNA sequencing applications, in the context of simplicity and high throughput, are the requirements both to amplify in vivo and then to purify the template prior to sequencing. PCR amplification as an alternative to in vivo DNA template preparation should minimize the need to remove cellular components that are amplified during traditional preparations. Furthermore, the compositions and reaction conditions of both PCR and dideoxy DNA sequencing are very similar. Therefore, we were able to develop an in situ sequencing protocol in which small aliquots of PCR-amplified DNA were added, with no intermediate purification, directly to dideoxy sequencing reactions. Double-stranded DNA templates generally require more sequencing skill than do single-stranded templates. Therefore, we developed a direct PCR-DNA sequencing procedure to transform all DNA templates into the single-stranded form, using asymmetric PCR priming (Gyllensten and Erlich 1988). The same sequencing protocol then could be used for both double- and single-stranded DNA. This general procedure for a dideoxycytidine sequencing reaction is as follows: bacteriophage λ DNA (Fig. 1; 1 ng) was heated on a thermal cycler (Perkin-Elmer Cetus) at 95°C for 3 minutes in the presence of 100 μl of PCR cocktail (1x PCR buffer and 0.2 M in deoxynucleoside triphosphates [both supplied by Perkin-Elmer Cetus], 1.0 pmole of PCR primer 639, and 50 pmoles of PCR primer 831), after being overlayed with light mineral oil. Then *Taq* polymerase (0.5 μl, 2.5 units [Perkin-Elmer Cetus]) was added, and 35 thermal cycles were performed (50°C for 2 min, 70°C for 3 min, plus an additional 20 sec each cycle, and 94°C for 1 min). Following amplification, 2 μl of this PCR solution was added to 3 μl of a solution containing 0.1 pmole fluorescein-labeled universal primer (FAM, Applied Biosystems) and 1x sequencing buffer (10 mM Tris-HCl [pH 8.5], 10 mM $MgCl_2$, 50 mM NaCl). After a 5-minute incubation at 55°C, 3.5 μl of T7 DNA polymerase solution (2 μl of d/dd CTP mix [United States Biomedical] and 1.5 μl of freshly prepared T7 mix: 3 units of T7 polymerase [United States Biochemical], 50 mM DTT, and 5 mM TE) was added and the solution was incubated at 37°C for 5 minutes. The three other sequencing reactions for dG, dA, and T were performed similarly, and they were pooled, ethanol precipitated, and loaded onto an Applied Biosystems 370 DNA Sequencer fitted with a 6% polyacrylamide gel.

With this protocol, we have achieved automated DNA sequence analysis from several targets of less than 2 kb in bacteriophage λ DNA, M13mp18(+), pGEM, and pUC, starting from 0.5 pg to 10 ng of DNA. Furthermore, we have eliminated

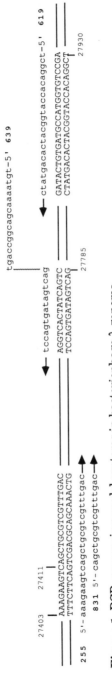

Figure 1 PCR-sequencing model systems in bacteriophage λ genome.

the need to synthesize and label template-specific sequencing primers by incorporating a universal primer sequence into the 5' end of the limiting PCR primer (Fig. 1). DNA targets not originally containing a universal primer sequence now can be sequenced using commercially available primers. Using this "universal" approach, we have sequenced a 352-bp portion of bacteriophage λ DNA (Fig. 1) in a single electrophoresis lane (Applied Biosystems 370A) and have produced >99% automatic base-calling accuracy over its entirety.

Direct Sequencing of Complex DNA Samples

We constructed a model system (Fig. 1) to develop direct sequencing of single copy genes using our universal PCR sequencing method. Human DNA (2 µg) purified from blood with a DNA extractor (Applied Biosystems) was spiked with 0.5 pg bacteriophage λ DNA (PCR control template, Perkin-Elmer Cetus). Because of the potential for PCR side products in complex DNA samples, we wanted to introduce an additional degree of selectivity for sequencing. This selectivity likely could be achieved by isolating the desired single-stranded PCR product from any side product by electrophoresis prior to sequencing. However, this approach would require prior knowledge of the electrophoretic mobility of the desired DNA product as well as the additional electrophoresis work. Alternatively, synthesis of a set of labeled primers (Gibbs et al. 1989) would give the added stringency for direct sequencing. However, to minimize labor we chose to do a two-step amplification (Fig. 1) using three PCR primers where 1% (1 µl) of the PCR products from primers 619 (3 pmoles) and 255 (20 pmoles) was added to a second PCR reaction (100 µl) containing a third primer 639 (1 pmole), which contained a universal primer sequence (–21, M13) and primer 255 (50 pmoles). Following a single dideoxy termination reaction (fluorescein sequencing primer) on 2 µl of this second PCR solution, a reliable signal-to-noise level was achieved (Applied Biosystems 370A). Comparison of these raw data, both to the data from a sample containing only human DNA as well as to a sample containing only bacteriophage λ DNA, revealed no detectable contribution from the human DNA to the noise level.

DISCUSSION

Successful direct PCR-sequencing from our model system of a complex DNA sample has demonstrated the feasibility of performing direct PCR-sequencing from extracted genomic DNA at the single copy level using this general method. This ability should facilitate rapid detection of certain genetic diseases (Gibbs et al. 1989). This two step approach (Fig. 1) also demonstrates the feasibility of a "PCR gene walking" strategy where sufficient sequence information already exists to perform the initial amplification.

A key component in developing our direct PCR sequencing protocol was the use of labeled sequencing primers. This labeling scheme renders the extension products due to false priming (for example, from the PCR primers) "invisible," since no label is associated with those products. Therefore, PCR primers do not have to be removed prior to sequencing. Alternatively, the use of labeled nucleoside trisphosphates during DNA sequencing likely would require isolation of the PCR product from the primers (potentially a difficult separation).

This requirement of using labeled sequencing primers motivated the development of a PCR procedure that incorporated the complement of a universal primer sequence into the amplified product (Scharf et al. 1986). Now, almost any PCR product under 2 kb can be sequenced by this approach without the need to synthesize fluorescently tagged sequence primers.

The developments in direct PCR-DNA sequencing procedures described above begin to illustrate the power of PCR applications in DNA sequencing. Our protocol for DNA template preparation and sequencing involves primarily repeated pipeting of small volumes and thermal cycling. Furthermore, we have sequenced double, single, circular and linear forms of DNA starting from 0.5 pg to 10 ng with the same protocol. Taking into consideration this universality, as well as the repetitive nature of this method conducive to automation, PCR-sequencing strategies might significantly contribute to many high-throughput and/or repetitive DNA sequencing projects. For example, we are trying to extend the use of this universal protocol to sequence directly from M13 plaques and bacterial colonies. PCR sequencing directly from colonies containing relatively small inserts should make shotgun cloning a more attractive strategy for ambitious sequencing projects.

ACKNOWLEDGMENTS
We thank Marilee Shaffer, Anne Wan, and Curt Becker for helpful discussions, and Nikki Griffey for manuscript preparation.

215

REFERENCES

Gibbs, R.A., P. Nguyen, L.J. McBride, S.M. Koepf, and C.T. Caskey. 1989. Identification of mutations leading to the Lesch-Nyhan syndrome by automated direct DNA sequencing of *in vitro* amplified cDNA. *Proc. Natl. Acad. Sci* **86:** (in press).

Gyllensten, U. and H.A. Erlich. 1988. Generation of single-stranded DNA by the polymerase chain reaction and its application to direct sequencing of the HLA-DQA locus. *Proc. Natl. Acad. Sci.* **85:** 7652.

Hood, L.E., M.W. Hunkapiller, and L.M. Smith. 1987. Automated DNA sequencing and analysis of the human genome. *Genomics* **1:** 201.

Innis, M.A., K.B. Myambo, D.H. Gelfand, and M.A.D. Brow. 1988. DNA sequencing with thermus aquaticus DNA polymerase, and direct sequencing of PCR-amplified DNA. *Proc. Natl. Acad. Sci.* **85:** 9436.

Mullis, K.B. and F.A. Faloona. 1987. Specific synthesis of DNA *in vitro* via a polymerase-catalyzed chain reaction. *Methods Enzymol.* **155:** 335.

Sanger, F., S. Nicklen, and A.R. Coulson. 1977. DNA sequencing with chain-terminating inhibitors. *Proc. Natl. Acad. Sci.* **74:** 5463.

Scharf, S.J., G.T. Horn, and H.A. Erlich. 1986. Direct cloning and sequence analysis of enzymatically amplified genomic sequences. *Science* **233:** 1076.

Smith, L.M., J.Z. Sanders, R.J. Kaiser, P. Hughes, C. Dodd, C. Connell, C. Heiner, S.B.H. Kent, and L.E. Hood. 1986. Fluorescence detection in automated DNA sequence analysis. *Nature* **321:** 674.

Stoflet, E.S., D.D. Koeberl, G. Sarkar, and S.S. Sommer. 1988. Genomic amplification with transcript sequencing. *Science* **239:** 491.

Rapid Analysis of T-cell Receptor Gene Structure and Expression

R.K. Wilson, D.H. Kono, D.M. Zaller, and L. Hood

Division of Biology, California Institute of Technology
Pasadena, California 91125

Identification and characterization of specific polymorphic or mutant alleles form the basis for studying human genetic diseases at the molecular level. Although several methods have been employed to detect and analyze these alleles, most require rigorous manipulations of a relatively large amount of target DNA (for review, see Caskey 1987; Landegren et al. 1988). The polymerase chain reaction (PCR) procedure (Saiki et al. 1985; Mullis and Faloona 1987), however, provides a means by which a single-copy gene can be specifically amplified to a level at which it may be detected by probe hybridization analysis or directly subcloned and subjected to nucleotide sequence analysis. Thus, PCR would seem to be a powerful method for detecting and characterizing disease-associated sequence mutations. Unfortunately, there are two drawbacks with this procedure that would appear to limit its general applicability. The first is the requirement for known sequences at both the 5' and 3' ends of the target region. This would limit the use of PCR solely to the study of previously characterized genes and homologous gene family members. The second drawback is the inherent infidelity of the thermostabile *Taq* DNA polymerase, which is used in the PCR procedure. Thus, when studying heterozygous or polymorphic alleles, several subclones must be sequenced to provide an accurate, unambiguous sequence. A major focus in our laboratory has been the development of techniques that facilitate rapid and efficient analysis of T-cell receptor (TCR) gene repertoire and expression in both normal and disease states. In this paper, we describe several strategies, which compensate for the two major drawbacks of the PCR procedure and facilitate the rapid analysis of TCR genes. These methods also are applicable to the study of other genes.

From total RNA:

I. known V gene segment

1. 1st strand synthesis using RT and poly(dT) primer
2. PCR using L and C primers.

II. Unknown V gene segment

1. 1st strand synthesis using RT and poly(dT) primer
2. tail with TdT and dGTP

3. PCR using N8-poly(dC) and C primers.

III. From amplified cDNA lysate:

IV. From germline DNA:

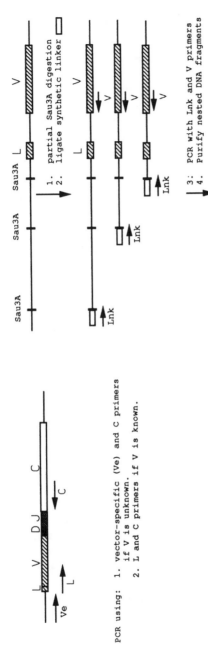

PCR using: 1. vector-specific (Ve) and C primers if V is unknown.
2. L and C primers if V is known.

1. partial Sau3A digestion
2. ligate synthetic linker □

3. PCR with Lnk and V primers
4. Purify nested DNA fragments

5. Sequence fragments using Lnk and V primers

Figure 1 Strategies for specific amplification of T-cell antigen receptor gene segment and flanking region sequences using PCR. Details for each strategy are given in the text.

The TCR is encoded by discontinuous variable (*V*), diversity (*D*), and joining (*J*) gene segments, which rearrange during T-cell development to form a functional *V* region gene (Wilson et al. 1988a). This gene is joined to constant (*C*) region sequences by posttranscriptional RNA processing. The 3' end of the *V* region, encoded by the *J* or *D-J* gene segments, functions as a junctional region. For every TCR allele, one rearranged *V-(D)-J* gene is expressed. Studies of TCR gene expression typically are performed by constructing genomic or cDNA libraries from T-cell lines or clones, screening these libraries with specific RNA or DNA probes, subcloning positive recombinant genomic or cDNA clones in an appropriate vector for nucleotide sequence analysis, and finally determining the nucleotide sequence of the subclone of interest. To circumvent these time-consuming procedures, we have developed the strategies outlined in Figure 1.

The first three strategies have been used to study the expressed TCR repertoire from normal and diseased animals. Two features are important for this. The first is to identify the unique *V* gene segment utilized, and the second is to sequence the junctional region to determine if the rearrangement of *V*, (*D*), and *J* segments is productive. The latter requires both in-frame *V-J* or *V-D-J* joining as well as the absence of termination codons within the junctional region. The fourth strategy is a technique with which the genomic flanking sequences of a *V* gene segment may be characterized. Rapid analyses of the subclones resulting from all of these strategies are facilitated by a high-throughput automated DNA sequencing system.

Strategy I: Identification of Unknown TCR *V* Gene Segments and *V-(D)-J* Junctional Sequences Using Oligonucleotides from Known *V* Gene Sequences

Compared with the immunoglobulin genes, the number of TCR *V* gene segment subfamilies in mice is relatively small, with 17 known for the β-chain and 17 known for the α-chain (Wilson et al. 1988a; E. Lai, unpubl.). In addition, several regions within the *V* gene segments are highly homologous between members of several different subfamilies. Using these two characteristics, we generated several *V* gene segment-specific oligonucleotides, which were used in combination with oligonucleotides antisense to C_α or C_β to screen first-strand cDNA from T-cell clones using the strategy outlined in Figure 1. PCR products then were subcloned in M13 and sequenced to identify un-equivocally the *V* gene segment used and to analyze the *V-(D)-J-C* junction to determine if the gene is productively rear-ranged. DNA sequencing is performed by the automated se-quencing facility, and a minimum of three subclones are ana-lyzed for each T-cell clone to compensate for the error rate ob-served with *Taq* DNA polymerase.

Strategy II: Identification of Unknown PCR *V* Gene Segments and *V-(D)-J* Junctional Sequences Using TdT Tailing

Strategies II and III have been used for identifying unknown *V* gene segments that do not cross-prime with any of the other *V* gene segment-specific oligonucleotides. Strategy II involves tailing of the first-strand cDNA with poly(dG) using terminal deoxynucleotide transferase (TdT) and dGTP. The product of the tailing reaction then is amplified using an antisense C_α or C_β oligonucleotide primer and a poly(dC) primer, which con-tains an additional 5' tail of eight nucleotides. The presence of the additional 5' nucleotides seems essential, since the use of poly(dC) alone yields a low amount of product. The resultant product again is subcloned in M13 and sequenced using the automated system. A major limitation of this procedure is the tendency of the poly(dC) oligonucleotide to bind other oligo-nucleotides that contain guanine triplets and to hybridize non-specifically to a multitude of other unknown sequences. This results in a background smear when the PCR product is sub-sequently analyzed on an agarose gel. Likely because of this problem, the yield of product using this technique is much lower than with Strategy I. However, full-length copies of *V*-region-specific sequences can be obtained faster and easier from a significantly lesser amount of RNA than with the con-ventional methods described above. After initial DNA sequence analysis, specific oligonucleotides can be prepared to screen other T-cell clones.

Strategy III: Characterization of TCR *V* Region Sequences from Preexisting cDNA Libraries

Strategy III is particularly useful for studying TCR gene ex-pression from preexisting cDNA libraries constructed in bac-teriophage λ vectors. For characterization of *V* gene segment utilization and junctional region sequences, *V*-region- and *C*-

region-specific oligonucleotides may be used for amplification as described above for total RNA. However, since the flanking vector sequences are known, one of two vector-specific primers (for either of two possible insert orientations) is used in place of the poly(dG) tailing strategy for characterization of unknown V gene segments. PCR is directly performed using a small amount of lysate from an amplified cDNA library, after heating at 70°C for a few minutes to disrupt the phage proteins. Following PCR, the amplification products are purified by agarose gel electrophoresis and subcloned in M13 for DNA sequence analysis. This strategy also is useful for amplification of DNA inserts from genomic and band libraries that have been constructed in bacteriophage λ vectors and can be used to circumvent library screening and subsequent subcloning tasks.

Strategy IV: Characterization of Unknown 5' -flanking Sequences Using Partial Restriction Digestion and Linker-specific Amplification

The identification of T-cell-specific promoter elements can be greatly facilitated by DNA sequence analysis of the 5' -flanking regions of TCR V regions. To simplify this task, we have developed a technique by which an unknown promoter region may be amplified, subcloned in M13, and sequenced. As outlined in Figure 1, this strategy allows amplification of 5' -flanking regions using only a single V-region-specific oligonucleotide primer. The target DNA is partially digested with the restriction enzyme Sau3AI and then ligated to synthetic linkers, which have a complementary four nucleotide overhang on one end. The linkers are left unphosphorylated so that they do not self-ligate and so that there is covalent attachment to only one strand of the DNA helix. PCR then is performed using a V-region-specific primer and a linker-specific primer. This results in the production of a set of overlapping DNA fragments that begin with a V region sequence and extend to various Sau3AI sites located in the 5' -flanking region. These amplified fragments then can be purified by agarose gel electrophoresis and sequenced either directly or following subcloning in M13. DNA sequence analysis of both strands is facilitated by using the V region and linker-specific oligonucleotides as sequencing primers. This strategy should be generally applicable for sequencing the promoter region of a gene corresponding to any known cDNA sequence.

High-throughput Automated DNA Sequence Analysis

Following PCR and M13 subcloning, amplified target DNAs from all of the strategies described above are rapidly sequenced using an automated high-throughput DNA sequencing system. As described previously (Wilson et al. 1988b), a modified robotic laboratory workstation is used to perform dideoxynucleotide DNA sequencing reactions for up to 24 DNA samples, using either radioisotopic or fluorescent labeling chemistries. If the fluorescent labeling chemistry is used, reaction products are analyzed on an automated fluorescent DNA sequencer, and the resulting nucleotide sequence data are directly stored on a computer hard disk. This DNA sequence data then are compared to a data base containing all previously characterized TCR gene sequences. Using this automated fluorescent system, over 100 subclones may be analyzed per week. With the automated, radioisotopic, DNA sequencing method, increased sample throughput is possible although laborious gel-loading and film-reading tasks are required. For detailed studies of TCR gene expression using the amplification strategies described above, a combination of both automated DNA sequencing methods was employed to process the large number of required subclones (D. Kono et al.; R. Wilson et al.; both in prep.).

REFERENCES

Caskey, C.T. 1987. Disease diagnosis by recombinant DNA methods. *Science* **236:** 1223.

Landegren, U., R. Kaiser, C.T. Caskey, and L. Hood. 1988. DNA diagnostics: Molecular techniques and automation. *Science* **242:** 229.

Mullis, K.B. and F.A. Faloona. 1987. Specific synthesis of DNA in vitro via a polymerase catalyzed chain reaction. *Methods Enzymol.* **155:** 335.

Saiki, R.K., S. Scharf, F. Faloona, K.B. Mullis, G.T. Horn, H.A. Erlich, and H. Kazazian. 1985. Enzymatic amplification of b-globin genomic sequences and restriction site analysis for diagnosis of sickle cell anemia. *Science* **230:** 1350.

Wilson, R.K., E. Lai, P. Concannon, R.K. Barth, and L.E. Hood. 1988a. Structure, organization and polymorphism of murine and human T-cell receptor α and β chain gene families. *Immunol. Rev.* **101:** 149.

Wilson, R.K., A.S. Yuen, S.M. Clark, C. Spence, P. Arakelian, and L.E. Hood. 1988b. Automation and dideoxynucleotide DNA sequencing reactions using a robotic workstation. *BioTechniques* **6:** 776.

Applications of PCR for the Construction of Vectors and the Isolation of Probes

P.J. de Jong, C. Chen, and J. Garnes

Biomedical Sciences Division, Lawrence Livermore
National Laboratory, Livermore, California 94550

DNA amplification by the polymerase chain reaction (PCR) offers new options for the site-specific introduction of changes in
target sequences available in plasmids. Multiple changes can
be introduced at two sites by the use of two PCR primers,
which differ from the corresponding sequences in the target
DNA by the presence of internal mismatches or 5' nonhybridizing tails. The primers can be used to amplify the region
bracketed by the two primers. The amplification product can
then be cloned into the plasmid to replace the original sequence. The ability to clone the PCR product as a substitute for
the original sequence depends on the presence of restriction
sites at the 5' end of the PCR product compatible with similar
sites in the plasmid. Although some of these changes can also
be made with synthetic restriction site adapters, this requires
far more screening work of the putative recombinant molecules.
We have used these procedures for the modification of a cosmid
vector and a yeast artificial chromosome (YAC) vector.

Design of the Lawrist5 Cosmid Vector
Cosmid vectors have the capacity to clone large fragments of
DNA in the size class of 35–45 kbp. The fragments to be cloned
are produced by partial digestion of genomic DNA with restriction enzymes that cut the genomic DNA rather frequently, generating fragments (in a total digest) with a size at least one order smaller than the desired partial digest fragment size. The
high frequency of occurrence of these sites is important to allow
the production of random genomic cosmid libraries. The enzyme used for the partial digest is therefore usually a four-
cutter, e.g., *Mbo*I (' GATC), which generates sticky ends compatible with a *Bam*HI (G' GATCC) cloning site in the vector.
The resulting hybrid sites in the recombinants are generally
not cleavable by *Bam*HI, and it is thus very unlikely that the

insert can be removed from the vector on a large genomic fragment. To permit the release of the entire insert from the cosmid, it would be desirable to have sites for "rare-cutting" restriction enzymes flanking the *Bam*HI cloning site. The *Sfi*I enzyme recognizes two 4-bp sequences separated by a random 5-mer: GGCCNNNN′NGGCC. The enzyme is known to cut mammalian DNA very rarely, resulting in fragments of average 149-kb size (Drmanac et al. 1986). Many cosmids from mammalian genomic DNA are therefore expected to have no *Sfi*I sites in the insert. The cohesive ends produced by *Sfi*I consist of random sequences of 3 bases with a 3′ overhang. A total of 64 different *Sfi*I cohesive ends is thus possible, corresponding with 32 possible *Sfi*I sites (two complementary overhangs per site). The *Sfi*I sites to be introduced into the vector may intentionally contain different cohesive ends. This then offers an option for differential labeling of the ends, useful for the mapping of insert restriction sites by a partial digest scheme (Smith and Birnstiel 1976). Furthermore, different sticky ends offer the possibility for the specific subcloning of sequences from the desired end and also allow transfer of the entire insert to different vectors (e.g., with a different replicon). In the case of such a transfer, only a single orientation of an insert relative to the vector will be possible, and a low background of nonrecombinants can be expected due to the mutual incompatibility of the two vector sticky ends. For these reasons, it was thus desirable to insert two different *Sfi*I sites in the vector (Lawrist4), adjacent to the cloning site, in a defined orientation.

The Lawrist4 vector is a derivative from LoristX (Speek et al. 1988). This is a λ-replicon vector with advanced features immediately flanking the cloning site, e.g., T7 and SP6 promotors and transcriptional terminators (Gibson et al. 1987). To not affect these features, the *Sfi*I sites were inserted flanking the cloning region (Fig. 1, ~400 bp, on an *Eco*RI fragment). Potentially, two different *Eco*RI/*Sfi*I/*Eco*RI adapters could have been synthesized and then inserted into the *Eco*RI sites. The insertions would have been random at both sites and many recombinants would have to be sequenced to detect the presence of the particular *Sfi*I sites in the desired place and polarity. PCR offers a more convenient option. Two PCR primers were synthesized with 3′ portions corresponding to the ends of the 400-bp *Eco*RI fragment. The nonhybridizing 5′ portions of the oligonucleotides contain the two different *Sfi*I sites, in the required

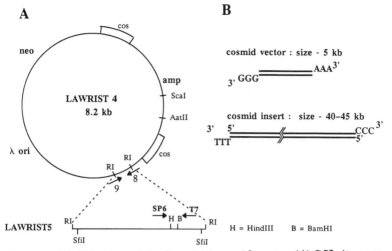

Figure 1 Modification of the Lawrist4 cosmid vector. (*A*) *Sfi*I sites are introduced into the vector using PCR primers (37-mers, nos. 8 and 9, with 18 bp identical to sequences in the Lawrist4 vector). The 5′ ends of the two primers contain different *Sfi*I sites (GGCCGAAA′ CGGCC and GGCCGGGG′ CGGCC) and *Eco*RI sites. After PCR amplification, an ~400-bp PCR product was obtained and used to replace the original *Eco*RI fragment in the Lawrist4 vector. The cloning sites in both cosmid vectors is *Bam*HI. (*B*) Cohesive end structures present at the *Sfi*I fragments from Lawrist5 cosmids.

orientation, and the *Eco*RI sites. Amplification starting with a few nanograms of the Lawrist4 vector generated the expected 400-bp product, which upon cleavage with *Eco*RI was ligated to the *Eco*RI replicon fragment. Putative recombinants were screened by restriction digests, and a clone containing the new 400-bp cloning region with the same polarity as the corresponding region of the previous construct was designated as the Lawrist5 vector. The vector has been applied to produce extensive cosmid libraries (>30-fold redundant) from several sorted human chromosomes (e.g., 19, 21, 22, and Y).

Modification of an Artificial Chromosome Vector

Recently, new vectors have been reported to enable the cloning of very large fragments of several 100 kbp in yeast (Burke et al. 1987). The recombinant molecules are propagated as YACs with vector sequences at the ends of the linear fragment. The vector sequences supply a centromere, two telomeres, selectable genes (*Trp*1, *Ura*3) and an autonomous replication sequence. In addition, one of the vector arms contains most of the pBR322 sequence, which facilitates the subcloning of probes from the corresponding end of the YAC. Sequences from the other end of the YAC insert are more difficult to subclone, requiring that a yeast genomic library be constructed in *Escherichia coli* and then screened with YAC vector sequences. The ability to retrieve sequences from the termini of large clone fragments is of obvious importance for (human) chromosome walking procedures to detect YAC clones partially overlapping with the previous clone. To facilitate the recovery of probes from both ends, we have constructed a derivative from the pYAC4 vector provided by D. Burke. First, subcloning would be facilitated if two different *Sfi*I sites with defined sticky ends would be present in both arms of the vector sequence. Because of the low frequency of *Sfi*I sites in yeast DNA, about 100 sites would be expected in the yeast clones. As discussed in the previous section, 64 possible sticky ends exist. Plasmid cloning vectors with *Sfi*I sites compatible with the sites in the modified YAC vector will allow rather specific subcloning from YAC sequences adjacent to vector sites. This will reduce the screening work needed to identify the subclone.

The modification of the vector was done by PCR, in a procedure analogous to the construction of the Lawrist5 cosmid vector (Fig. 2). Two long PCR primers, a 67-mer (no. 56) and a 70-mer (no. 55), were synthesized. These primers have 13-bp and 16-bp regions at their 3′ ends, respectively, which correspond to sequences flanking the *SUP4*-ochre (*SUP4*-o) gene in pYAC4. The *Eco*RI cloning site in the YAC vector is located in the *SUP4*-o intron sequence. The long 5′ portions of the primers contain a series of features, including the two *Sfi*I sites. The amplification product generated with these two PCR primers had the expected approximately 300-bp size. The product could be cloned in the pYAC4 vector because of the *Cla*I and *Sal*I sites present at the 5′ end of the two PCR primers, to substitute for the original *SUP4*-o gene. The resulting vector is termed pYAC4-LL1.

Isolation of Probes from the Ends of YAC Inserts by Inverted PCR

Both PCR primers, used for the modified YAC vector, also include a T3 promotor sequence, directed away from the insert, and a *Sna*BI site (TAC′ GTA). The purpose of these sequences is to allow the amplification of end probes from YAC inserts by

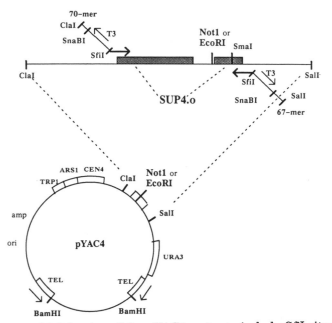

Figure 2 Modification of the pYAC4 vector to include *Sfi*I sites and T3 promotors. Two oligonucleotide primers were synthesized, which had 16 bp and 13 bp at the 3′ end in common with vector sequences, upstream of and downstream from the yeast *SUP4*-o gene. The oligonucleotides were used to generate a PCR product of the expected ~300-bp size. The PCR product was cleaved with *Cla*I and *Sal*I and then inserted into pYAC4 cleaved with the same enzymes. This generated the new vector, pYAC4-LL1.

PCR (Fig. 3). *Sna*BI is a rare cutting enzyme in human DNA due to the presence of a CpG dinucleotide sequence in the recognition site. Yeast genomic DNA from a YAC clone can be cut to completion in a double digest with *Sna*BI and a frequently cutting enzyme, e.g., *Rsa*I or *Alu*I also generating blunt ends. Two junction fragments are thus released that contain both vector and insert sequences. Ligation of the digest fragments at low DNA concentration will primarily generate circular products because of a new (artificial) junction of vector sequences (*Sna*BI site) to insert sequences (*Rsa*I or *Alu*I site). The junction circles can then be opened again with *Sfi*I, producing linear fragments with YAC vector sequences present at both ends and with the T3 promotor now immediately adjacent to the insert sequences. This procedure will likely allow the PCR amplification of terminal YAC sequences. One of the prim-

229

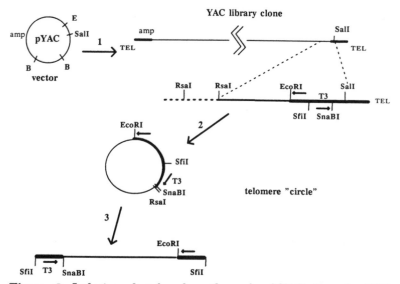

Figure 3 Isolation of probes from the ends of YAC clones by PCR using circular intermediates. (1) YAC clones are prepared by ligation of the *Bam*HI/*Eco*RI fragments from the pYAC4-LL1 vector to partial *Eco*RI digest fragments from genomic (human) DNA. (2) DNA from a specific YAC clone is isolated and cleaved with *Rsa*I and *Sna*BI. At low DNA concentration, the DNA is subsequently ligated to generate primarily circular product. (3) Circles containing vector/insert junctions are linearized and then used as a source for PCR using a primer adjacent to the *Eco*RI cloning site in combination with the T3 promotor oligonucleotide.

ers useful for both YAC ends is the T3 promotor sequence. The other PCR primer corresponds to vector sequences next to the *Eco*RI cloning site. The choice of this primer determines which end will amplify. Furthermore, by selecting different restriction enzymes or by the use of partial digest conditions, a series of circular intermediates can be generated that all contain the same amount of vector sequence (from the *Eco*RI cloning site up to the *Sna*BI site) and different amounts of YAC insert sequences. Since the T3 promotor is immediately adjacent to the *Sna*BI site, this means that the T3 promotor is placed next to insert sequences at variable distances from the *Eco*RI cloning site. This feature might be important to avoid dispersed repetitive sequences being transcribed into RNA probes. Finally, it should be emphasized that the PCR amplification of YAC terminal sequences is still untested. Recently, two publications have appeared describing different applications of circular intermediates for PCR reactions (Ochman et al. 1988; Triagla et al. 1988).

ACKNOWLEDGMENT

This work was performed under the auspices of the U.S. Department of Energy by the Lawrence Livermore National Laboratory under contract no. W-7405-ENG-48.

REFERENCES

Burke, D.T., G.F. Carle, and M.V. Olsen. 1987. Cloning of large exogenous DNA into yeast by means of artificial chromosome vectors. *Science* **236:** 806.

Drmanac, R., N. Petrovic, and R. Crkvenjakov. 1986. A calculation of fragment lengths obtainable from human DNA with 78 restriction enzymes: An aid for cloning and mapping. *Nucleic Acids Res.* **14:** 4691.

Gibson, T.J., A.R. Coulson, J.E. Sulston, and P.F.R. Little. 1987. Lorist2, a cosmid vector with transcriptional terminators insulating vector genes from interference by promotors within the insert: Effect on DNA yield and cloned insert frequency. *Gene* **53:** 275.

Ochman, H., A.S. Gerber, and D.L. Hartl. 1988. Genetic application of an inverse polymerase chain reaction. *Genetics* **120:** 621.

Smith, M.O. and M.L. Birnstiel. 1976. A simple method for DNA restriction site mapping. *Nucleic Acids Res.* **3:** 3287.

Speek, M., J.W. Raff, K. Harrison-Lavoie, P.F.R. Little, and D.M. Glover. 1988. Smart2, a cosmid vector with the phage lambda origin for both systematic chromosome walking and P-element-mediated gene transfer in *Drosophila*. *Gene* **64:** 173.

Triagla, T., M.G. Peterson, and D.J. Kemp. 1988. A procedure for in vitro amplification of DNA segments that lie outside the boundaries of known sequences. *Nucleic Acids Res.* **16:** 8186.

Allele-specific Amplification Reactions: LAR and ASPCR

D.Y. Wu, L. Ugozzoli, B.K. Pal, and R.B. Wallace

Beckman Research Institute of the City of Hope
Duarte, California 91010

Bacteriophage T4-induced DNA ligase catalyzes the ATP-dependent phosphodiester bond formation between the 5′ phosphoryl and the 3′ hydroxyl groups of adjacent DNA strands. The mechanism involves initially the formation of an enzyme-adenylate complex, followed by the transfer of the adenylate from the enzyme to the 5′ phosphoryl group of the substrate and finally the formation of the phosphodiester bond between the adjacent DNA substrates. In this process, one molecule of ATP is cleaved into AMP and inorganic pyrophosphate (PP_i) with the formation of one phosphodiester bond.

Duplex DNA molecules with either cohesive ends or blunt ends can serve as ligation substrates, although cohesive-end ligation is far more efficient. However, single-stranded DNA is not a substrate. In addition, T4 DNA ligase can repair single-stranded breaks (nicks) in double-stranded DNA molecules.

Specificity of T4 DNA Ligase Nick-closing Activity

In the past, most studies have focused on the ability of T4 DNA ligase to catalyze blunt-end and cohesive-end joining of synthetic polynucleotide or plasmid substrates. Comparatively little is known about the nick-closing activity of this enzyme. We have designed a model for the nick-closing activity based on the joining of adjacent, short, synthetic oligonucleotides on cDNA templates. A series of synthetic oligonucleotide pairs (8 and 14 nucleotides long) that are complementary to either the normal (β^A) or the sickle cell (β^S) alleles of the human β-globin gene at the sequence surrounding the sickle cell mutation is shown in Figure 1.

T4 DNA ligase effectively joins two adjacent, short, synthetic oligonucleotides, on complementary oligonucleotide, plasmid, and genomic DNA templates. When a single base-pair mismatch exists at either side of the ligation junction, the efficien-

```
ONS2:3'-T G A G G A C A

ONA2:3'-T G A G G A C T

                    ON1:3'-C C T C T T C A G A C G G C
(+)5'...G C A C C T G ACTCCTGAGGAGAAGTCTGCCG T...3'
                               *
(-)3'...C GTGGACTGAGGACT CCTCTTCA G A C G G C A...5'
  ONA3:5'-C A C C T G A C T C C T G A

  ONS3:5'-C A C C T G A C T C C T G T

                    ON4:5'-G G A G A A G T
```

Figure 1 The sequence of the region of the normal human β-globin gene (β^A) to which the oligo substrates are complementary is shown. ON1 and ON2 (either ONA2 or ONS2) are complementary to the coding plus strand in the shaded region. ONA2 is completely complementary to the sequence shown, and ONS2 is complementary to the sickle cell allele (β^S). The 5' nucleotide of ONS2 forms an A-A mismatch with the β^A allele. ON3 (either ONA3 or ONS3) and ON4 are complementary to the noncoding minus strand in the shaded region. ONA3, like ONA2, is completely complementary to the β^S gene, and ONS3 is complementary to the β^S allele. The 3' nucleotide of ONS3 forms a T-T mismatch with the β^A allele. The ligation junctions of the adjacent ON1/ON2 substrates and ON3/ON4 substrates are directly over the base pair of interest (indicated by an asterisk), the position of the sickle cell (A→T) transversion mutation.

cy of the enzyme to ligate the two oligonucleotides decreases. Mismatch ligation is approximately fivefold greater if the mismatch occurs at the 3' side rather than at the 5' side of the junction. During mismatch ligation, the 5' adenylate of the 3' oligonucleotide accumulates in the reaction. The level of the adenylate formation correlates closely with the level of the mismatch ligation. Both mismatch ligation and adenylate formation are suppressed at elevated temperatures and in the presence of 200 mM NaCl or 2–5 mM spermidine. The apparent K_m for the oligonucleotide template in the absence of salt is 0.05 µM, whereas the K_m increases to 0.2 µM in the presence of 200 mM NaCl. Using the oligonucleotide pairs complementary to the β-globin gene at the sequence surrounding the single base-pair mutation responsible for sickle cell anemia (see above), the ability of T4 DNA ligase to ligate the oligonucleotides can be used to discriminate the two alleles. The highly specific nature of the nick-closing reaction, therefore, can be used to distinguish two DNA templates differing by a single nucleotide.

Ligation Amplification Reaction

Amplification of the products of ligation is possible. The product is increased by either linear or exponential amplification using sequential rounds of template-dependent ligation. In the case of linear amplification, a single pair of oligonucleotides is ligated, the reaction is heated to dissociate the ligation product, and an additional round of ligation is performed. After n rounds, there is a $(1+X) \times n$-fold amplification of product where X is the efficiency of the ligation reaction. Exponential amplification utilizes two pairs of oligonucleotides, one complementary to the upper strand and one to the lower strand of a target sequence. The products of the ligation reaction serve as templates for subsequent rounds of ligation. In this case, there is $(1+X)^{(n-1)}$-fold amplification of product after n rounds. A single base-pair mismatch between the annealed oligonucleotides and the template prevents ligation, thus allowing the distinction of single base-pair differences between DNA templates. At high template concentrations, the ligation reaction has an efficiency approaching 100%. In this paper, we demonstrate the use of the ligation amplification reaction (LAR) to distinguish the normal from the sickle cell allele of the human β-globin gene. LAR can also be used as an allele-specific detection system for polymerase-chain-reaction (PCR)-enriched DNA sequences.

Allele-specific Polymerase Chain Reaction

Recently, we have determined that PCR itself can be done in an allele-specific manner. This has led to a rapid, nonradioactive approach to the diagnosis of sickle cell anemia based on allele-specific PCR (ASPCR). This method allows the direct detection of the normal or the sickle cell allele in genomic DNA without additional steps of probe hybridization, ligation, or restriction enzyme cleavage. Two allele-specific oligonucleotide primers, one specific for the sickle cell allele and one for the normal allele, were used in PCR together with another primer complementary to both alleles. The allele-specific primers differed from each other in their terminal 3' nucleotide. Under the proper annealing temperature and PCR conditions, these primers only directed amplification on their complementary allele. In a single, blind study of DNA samples from 12 individuals, this method correctly and unambiguously allowed for the determination of the genotypes with no false negatives or positives. When combined with appropriate labeling techniques, this method promises to be a powerful approach for genetic disease diagnosis, carrier screening, HLA typing, human gene mapping, forensics, and paternity testing.

ACKNOWLEDGMENTS
This work was supported by grant DCB-8515365 from the National Science Foundation (R.B.W.). D.Y.W. is an M.D./Ph.D. candidate at Loma Linda University. R.B.W. is a member of the Cancer Center of the City of Hope (National Institutes of Health grant CA-33572).

The Polymerase Chain Reaction: Why It Works

K.B. Mullis

Private Consultant
6767 Neptune Avenue, Apt. 4
La Jolla, California 92037

The polymerase chain reaction (PCR) (Mullis et al. 1986; Mullis and Faloona 1987; Mullis 1987) is suddenly a standard laboratory technique. An enzymatic reaction, as simple to perform as it is intellectually satisfying to contemplate, the PCR solves two of the more universal problems in the chemistry of natural nucleic acids. It allows for the physical separation of any particular sequence of interest from its context, and then it provides an in vitro amplification of this sequence, virtually without limit. How is it that a simple arrangement of two oligodeoxynucleotides, four deoxynucleoside triphosphates, and a DNA polymerase can produce such a result? The surprising robustness of PCR, the fact that it is difficult to find situations in which it will not work, derives from its fortuitous combination of three familiar phenomena, each of which is intrinsically powerful.

The first of these is the impressive ability of almost all oligodeoxynucleotides, under a wide range of conditions, to bind tightly and specifically to their complementary nucleic acid sequences. These short single-stranded DNA molecules are able to sort through, as it were, hundreds of thousands of likely binding sites, discriminating easily between a perfect fit and an almost perfect fit, binding tightly to the perfect complement, and at practical concentrations and temperatures, accomplishing this in milliseconds. A respectable shot is 1 in 100,000. If the human genome contained only a few hundred thousand base pairs, the selectivity of an oligodeoxynucleotide alone would enable most conceivable discriminations to be made readily. The human genome, however, consists of about 3.3 x 10^9 bp, and as had been pretty well demonstrated (Wallace et al. 1986) prior to the emergence of PCR, something in addition to oligodeoxynucleotide specificity is necessary to make the specific sampling of human DNA sequence variation possible. Theoretically, the information contained in a 17-base-long oligodeoxynucleotide is sufficient to define a precise location on a random sequence the size of the human genome ($4^{17} = 1.72 \times 10^{10}$ is greater than 6.6×10^9, which is the number of nucleotides available for hybridization when the human genome is denatured). However, absolute specificity is not obtained in practice even at high temperatures, and there are good reasons why not. Most of them have to do with the fact that if something else can happen, it does. Using longer oligodeoxynucleotides and/or involved and/or secret protocols doesn't markedly improve the situation. Thus, oligodeoxynucleotides provide a good starting point for procedures that must address specific sequences, but they are not in themselves sufficient.

The second familiar phenomenon is illustrated by the notion that the probability for the occurrence of a compound action is the product of the individual probabilities for the occurrence of each of its components. What has this got to do with PCR? Everything. This is also the critical feature of "sandwich" format immunochemical assays. In both cases, "correct" binding events, which are very probable, are unavoidably associated with a very much larger number of "incorrect" binding events. The "incorrect" events are individually improbable, but the whole hoard of them collectively constitute an unacceptably high background. PCR, like the sandwich assays, succeeds by strictly requiring that a final signal can only be generated by the coordinated action of two binding events. In the case of PCR, a fragment is only amplified if two oligodeoxynucleotides are bound to the single-stranded DNA templates in such a way that the polymerase extension product of the one contains a binding site for the other. The probability for this in the presence of the intended DNA target is the product of the probabilities of the individual oligodeoxynucleotides finding their binding sites on the target DNA strands. Under appropriate conditions of concentration, time, and temperature, the individual binding probabilities are close to 1.0, and hence, so is their product. Therefore, during every cycle of PCR, there is a high probability that a new copy of the target is produced from each existing copy. However, for any particular nontarget stretch of DNA to be amplified, 2 probabilities far less than 1 (maybe 10^{-5}, that being the approximate probability that a single oligodeoxynucleotide will bind to a random DNA sequence) have to be multiplied, and the probability of the compound action is therefore very low (maybe 10^{-10}). The statistical treatment here is incomplete but should be suggestive of the way the PCR improves remarkably on the degree of specificity that can ultimately be obtained from the binding of oligodeoxynucleotides to targets on complex DNA.

The third familiar phenomenon embodied in PCR relates to the properties of a chain reaction itself. The PCR is a sort of deoxyribonuclear bomb. Exponential growth is obtained by repeated copying of templates into new templates. This undoubtedly imbues the PCR with much of its aforementioned robustness. PCR amplifications are in this way also similar to gene pools. Both are harder to terminate with a single blow because they are carried in a population that is continually dividing, wherein the malfunction of an individual is not fatal for the pool. A linear process can be cut off at a single source, not so with a chain reaction. The process of division provides more and more growing points. Exponential growth of any kind is difficult to step on, but exponential growth, which is a result of a chain reaction in which products become reactants, is a particularly viable beast. Exponential growth is also important from a practical point of view regarding the time required for a given level of amplification. However, this is only a matter of convenience. If we had to wait, we could.

Amplification is the most striking result of PCR. It has allowed us to look at single molecules and to look in biological samples for very rare occurrences. However, amplification aside, the ability to prepare pure DNA sequences easily is starting to have an impact on the way molecular biology is done and on the type of experiments that a molecular biologist will consider. Relatively pure fragments of DNA, the ends of which are not dependent on the fortuitous presence of restriction enzyme sites, are now accessible, and these can be modified easily. The fact that the 5' end of the oligodeoxynucleotides is not critical to the PCR reaction itself allows the investigator the option to insert functional DNA sequences or nonnucleic acid substituents onto the ends of the targeted fragment. Thus, restriction enzyme sites, RNA polymerase promoter sequences, organic chemical handles for capturing or detecting a sequence, and so forth, can be easily appended as the fragment of interest is being isolated. The problem of getting flanking sequences by a PCR mechanism is one that many investigators have thought about for a long time, and finally this year it was solved in a particularly nice way and apparently in more than one laboratory simultaneously (Ochman et al. 1988; Triglia et al. 1988). The trick is to circularize the target DNA and turn the primers for a known region around backwards, thereby amplifying the flanks. In addition to other uses, this clearly will be a real boon for sequencing the human genome.

Some Practical Comments

The precise conditions under which PCR always works best are not known, and it seems reasonable to assume that no such standard conditions will be discovered. After all, people use PCR for very different purposes. The 1000-fold amplification of a plasmid-born fragment starting from 100 fmoles is different from the 10^{11}-fold amplification of several molecules up to 100 fmoles (Kim and Smithies 1988), and amplifying a 58-bp fragment is not the same as amplifying a 10-kb fragment (Jeffreys et al. 1988). There are some things, however, that are generally true and which may be useful to keep in mind when setting up a particular reaction.

Enzyme and Primer Concentration

One unit of the polymerase from *Thermus aquaticus* (*Taq*) is about 50 fmoles of the protein (Innis et al. 1988). This compares with 500 fmoles/unit for Klenow. When using 3 units of *Taq* in a 100-μl reaction, the enzyme is 1.5 nM. If you are amplifying a fragment up to the 10 nM level, which is where a 5–10-μl aliquot becomes readily visible on an ethidium-stained gel (depending on the number of base pairs in the fragment), the fact that the enzyme will become limiting at some point should be considered. At fragment concentrations greater than 1.5 nM, some molecules of the enzyme are being asked to extend two fragments per cycle.

The success of the PCR reaction depends on the kinetic advantage that high concentrations of primers have over relatively low concentrations of product strands, which at equilibrium would rehybridize with each other and displace the primers. This advantage is lost if sufficient enzyme is not available during the moments immediately after the primers have hybridized. Likewise, the primer concentration must be high enough to begin with so that the primers do indeed get there first. Primer concentrations less than 100-fold greater than expected product concentrations would be asking a lot more than oligonucleotides can probably provide. The relative importance of these parameters is hard to dissect from the available data, but the most impressive amplifications seen so far, in terms of very-high-gain amplifications (Kim and Smithies 1988), ampli-

240

fication of one or several fragments to high levels (Mullis and Faloona 1987; Chamberlain et al. 1988), or amplification of very large fragments (Jeffreys et al. 1988), have all used large amounts of enzyme (5–15 units/100 μl) and primers between 1 and 10 μM. For amplifications requiring many cycles, enzyme has been added more than once.

Reaction Time

This can be very short for fragments in the 50–200 bp range; I suggest trying no time at all at first. Cycle back and forth as fast as the cycler will allow between 60°C and 95°C. Compare this with holding at 60°C or stopping at 75°C for 2 or more minutes. For long fragments, a 10- or 20-minute wait may be necessary. The temperature optimum for the enzyme itself on an M13 template is about 74°C.

From the molecular weight of 94,000 and specific activity of approximately 250,000 units/mg (D.H. Gelfand and S. Stoffel, in prep.), one can calculate a turnover number or minimum extension rate at 74°C of 131 nucleotides/sec. The extension rate determined from labeled primers (Innis et al. 1988) is 60–120 nucleotides/sec at 80°C, but the precise relevance of these numbers to PCR reactions is not exactly clear. It has been observed in several laboratories that longer fragments can require 10–20 minutes per extension cycle to achieve optimum yields, which still is lower than short fragment yields. If you are going to be amplifying the same fragment over and over again, you should do some experiments to determine the best times.

Magnesium, dNTPs, and Melting Temperature

Every double-stranded DNA does not denature at the same temperature, not even close. It is not known whether complete denaturation is necessary for amplification, but it is known that a number of amplifications have failed because of insufficient heating. For the time being, we are stuck with an enzyme that cannot be boiled, which is a shame. We are using this particular enzyme not because it is the best possible enzyme for the job, but because I somewhat arbitrarily picked it from the four or five thermophilic enzymes that were described in the literature by 1984. After it was purified in D. Gelfand's laboratory (D.H. Gelfand and S. Stoffel, in prep.), and we found that it worked okay for PCR, nobody saw fit to purify any others. Someone should, but in the meantime, we have to deal with an enzyme that begins to fall apart seriously at about

95°C, and in some cases that causes trouble. If the fragment is particularly hard to melt, a higher temperature, lower salts, or a little dimethylsulfoxide (DMSO) might all help raise the yield. However, higher temperatures are out, and the enzyme requires (W. Sutherland [Eastman Kodak], pers. comm.) at least 1 mM free magnesium, meaning concentrations of magnesium 1 mM greater than the total concentration of all the dNTPs (which have a taste for magnesium themselves). So, you're between a rock and a hard place again. Some people lower the magnesium anyhow, and consequently, they must also lower the dNTPs. By running under conditions that are near marginal for magnesium, they are risking variable results if samples are introduced that contain various levels of magnesium. Sodium and/or potassium are more or less ornamental components of some recipes that are not necessary for the reaction and can be left out entirely without hurting anything. Dropping the monovalent ion concentration by 50 mM is equivalent to dropping the magnesium concentration by 5 mM in terms of destabilizing double-stranded DNA, and 10% DMSO has been added from time to time without disastrous side effects. None of these parameters have been studied sufficiently by anybody to raise the level of this discussion higher than just the local gossip on the PCR circuit. The reason for this is perhaps that most of the time no such efforts are necessary. The fragments melt easily below 95°C, and the reactions work fine under any reasonable conditions.

Out of a natural laziness, I always start with the easiest possible protocol and work from there. Better yet, I suggest that someone else start from there, and I come back in a month to see how things worked out. The easiest possible protocol is one where the numbers are all round and even powers of ten, so I start from the following: 100 mM Tris-HCl (pH 9.0, measured at whatever the room temperature happens to be); 10 mM $MgCl_2$; 1–10 μM oligodeoxynucleotides; 1 mM dNTPs; and 10 units of Taq/100 μl reaction. Remembering that long DNA molecules are more resistant to heat than relatively small PCR fragments, thoroughly denature the DNA sample prior to adding the enzyme by immersing the reaction tube in boiling water for 1 minute. Add the enzyme and start cycling back and forth between 65°C and 95°C as fast as possible. Begin with 10 fmoles of a cloned target, run ten cycles, put 10 μl from the last cycle on an agarose gel with 0.5 μg/ml of ethidium bromide in it, and look for a band after a few minutes. If it works, then try more cycles from less sample. If it does not, vary the parameters until it does. Get it right at this level before going on to something more demanding, and don't do anything you don't have to do because every step you add to your protocol is likely to ruin it.

The recipes used by the two groups at this meeting, who presented what I thought were the most impressive PCR results to date (Kim and Smithies 1988; and Caskey's Baylor Boys, Chamberlain et al. 1988), did not employ round numbers. They both use 67 mM Tris (pH 8.8), 16.7 mM ammonium sulfate, 6.7 mM magnesium chloride, bovine serum albumin, and other less cryptic materials, as can be found in their papers.

REFERENCES

Chamberlain, J.S., R.A. Gibbs, J.E. Ranier, P.N. Nguyen, and C.T. Caskey. 1988. Deletion screening of the Duchenne muscular dystrophy locus via multiplex DNA amplification. *Nucleic Acids Res.* **16:** 11141.

Innis, M.A., D.H. Gelfand, K.B. Myambo, and M.D. Brow. 1988. DNA sequencing with *Thermus aquaticus* DNA polymerase and direct sequencing of polymerase chain reaction amplified DNA. *Proc. Natl. Acad. Sci.* **85:** 9436.

Jeffreys, A.J., V. Wilson, R. Neumann, and J. Keyte. 1988. Amplification of human minisatellites by the polymerase chain reaction: Towards DNA fingerprinting of single cells. *Nucleic Acids Res.* **16:** 10953.

Kim, H.-S. and O. Smithies. 1988. Recombinant fragment assay for gene targetting based on polymerase chain reaction. *Nucleic Acids Res.* **16:** 8887.

Mullis, K.B. 1987. Process for amplifying nucleic acid sequences. *US Patent 4,683,202*. Filed 10-25-85; Issued 7-28-87.

Mullis, K.B. and F. Faloona. 1987. Specific synthesis of DNA in vitro via a polymerase-catalyzed chain reaction. *Methods Enzymol.* **155:** 335.

Mullis, K., F. Faloona, S. Scharf, R. Saiki, G. Horn, and H. Erlich. 1986. Specific enzymatic amplification of DNA in vitro: The polymerase chain reaction. *Cold Spring Harbor Symp. Quant. Biology.* **51:** 263.

Ochman, H., A.S. Gerber, and D.L. Hartl. 1988. Genetic applications of an inverse polymerase chain reaction. *Genetics* **120:** 621.

Triglia, T., M.G. Peterson, and D.J. Kemp. 1988. A procedure for in vitro amplification of DNA segments that lie outside the boundaries of known sequences. *Nucleic Acids Res.* **16:** 8186.

Wallace, R.B., L.D. Petz, and P.Y. Yam. 1986. Application of synthetic DNA probes to the analysis of DNA sequence varients in man. *Cold Spring Harbor Symp. Quant. Biol.* **51:** 257.